CCEA
GCSE
PHYSICS
THIRD EDITION

Roy White
Frank McCauley

HODDER
EDUCATION
AN HACHETTE UK COMPANY

Photo credits

p.1 © Mikhail Mandrygin/123RF; **p.12** © Germanskydiver/Shutterstock; **p.20** © Germanskydiver/Shutterstock; **p.24** © Gary Hawkins/Rex Features; **p.33** © Gary Hawkins/ Rex Features; **p.39** © Senohrabek/Shutterstock; **p. 46** © Diyana Dimitrova/123RF; **p.62** © Diyana Dimitrova/123RF; **p.63** © iurii/Shutterstock; **p.64** © iurii/Shutterstock; **p.74** © David Parker/Science Photo Library; **p.86** br © Mediscan/Alamy, tl © Chris Priest/Science Photo Library, bl © Hank Morgan/Science Photo Library; **p.93** © David Parker/Science Photo Library; **p.97** © Darren Brode/Fotolia; **p.105** © Darren Brode/Fotolia; **p.110** © Piotr Krzeslak/Shutterstock; **p.132** © PhotosIndia.com LLC/123RF; **p.156** © Chakrapong Worathat/123RF; **p.157** tl © Chakrapong Worathat/123RF, tr © Konstantin Gushcha/Fotolia, bl © misterryba/Fotolia, bm © Jaraslow Grudzinski/Fotolia, br © Jamie Kingham/Getty Images; **p.165** © NG Images/Alamy; **p.184** © Reinhold Wittich/123RF; **p.189** © The Print Collector/Alamy; **p.194** © Reinhold Wittich/123RF.

Text acknowledgements

CCEA Science: Double Award, Unit 1, HT, Feb 2013: Q8 (**p.11**, Q13); Q9: (**p.62**, Q11); Double Award, Unit 1, FT, Feb 2013: Q10c (**p.38**, Q13a); Q7 (**p.38**, Q13b); Double Award, Unit 1, HT, May 2013: Q8 (**p.22**, Q1); Q4 (**p.44**, Q4); Double Award, Unit 1, HT, Feb 2008: Q10 (**p.22**, Q2); Double Award, Unit 1, HT, Feb 2009: Q11 (**p.22**, Q3); Q4 (**p.35**, Q5); CCEA Science: Double Award, Unit 1, HT, Nov 2006: Q11 (**p.22**, Q4); Double Award, Unit 1, HT, Nov 2007: Q11 (**p.22**, Q5a); Double Award, Unit 1, HT, Feb 2007: Q11 (**p.23**, Q8); Double Award, Unit 1, HT, Feb 2014: Q8a: (**p.22**, Q5b), Q7 (**p.62**, Q12); Double Award, Unit 1, HT, June 2013: Q1b (**p.36**, Q9a); Double Award, Unit 1, HT, May 2012: Q10b (**p.36**, Q9B); Q4 (**p.44**, Q3); Physics, Unit 1, HT, June 2012: Q5a and b (**p.37**, Q12); Physics, Unit 1, HT, June 2012:Q1 (**p.62**, Q13); Physics, Unit 1, HT, June 2014: Q3a (**p.44**, Q5a); Double Award, Unit 1, HT, Nov 2012: Q4 (**p.95**, Q3); Double Award, HT, May 2014: Q8 (**p.36**, Q8); Q9 (**p.62**, Q10)

Although every effort has been made to ensure that website addresses are correct at time of going to press, Hodder Education cannot be held responsible for the content of any website mentioned in this book. It is sometimes possible to find a relocated web page by typing in the address of the home page for a website in the URL window of your browser.

Hachette UK's policy is to use papers that are natural, renewable and recyclable products and made from wood grown in well-managed forests and other controlled sources. The logging and manufacturing processes are expected to conform to the environmental regulations of the country of origin.

Orders: please contact Hachette UK Distribution, Hely Hutchinson Centre, Milton Road, Didcot, Oxfordshire, OX11 7HH. Telephone: +44 (0)1235 827827. Email education@hachette.co.uk Lines are open from 9 a.m. to 5 p.m., Monday to Friday. You can also order through our website: www.hoddereducation.co.uk.

© Roy White and Frank McCauley

First edition published 2003

Second edition published 2011

Third edition published 2017 by
Hodder Education,
An Hachette UK Company
Carmelite House
50 Victoria Embankment
London EC4Y 0DZ

Impression number 10 9 8 7 6 5 4
Year 2022

Cover photo © Stocktrek Images, Inc. / Alamy Stock Photo

Illustrations by Elektra Media Ltd

Typeset by Elektra Media Ltd

Printed by CPI Group (UK) Ltd, Croydon CR0 4YY

A catalogue record for this title is available from the British Library.

ISBN 9781471892172

CONTENTS

How to get the most from this book

UNIT 1

UNIT 2

HOW TO GET THE MOST FROM THIS BOOK

Welcome to the CCEA GCSE Physics Student Book.

This book covers all of the Foundation and Higher-tier content for the 2017 CCEA GCSE Physics specification.

The following features have been included to help you get the most from this book.

Specification points

Check that you are covering all the required content for your course, with specification references and a brief overview of each chapter.

Tip

These highlight important facts, common misconceptions and signpost you towards other relevant chapters. They also offer useful ideas for remembering difficult topics.

Show you can

Complete the Show you can tasks to prove that you are confident in your understanding of each topic.

Test yourself

These short questions, found throughout each chapter, allow you to check your understanding as you progress through a topic.

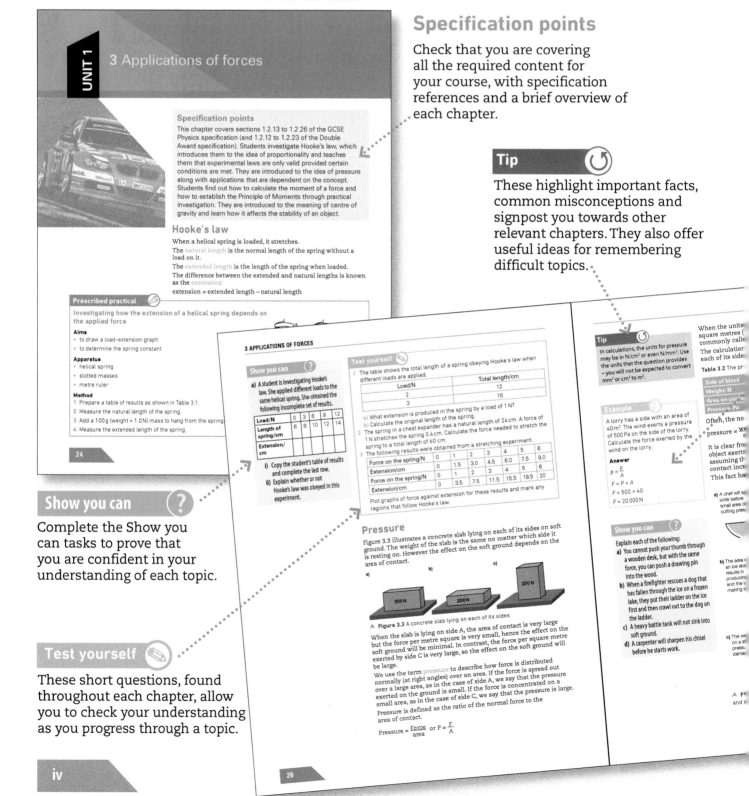

UNIT 1
3 Applications of forces

Specification points

This chapter covers sections 1.2.13 to 1.2.26 of the GCSE Physics specification (and 1.2.12 to 1.2.23 of the Double Award specification). Students investigate Hooke's law, which introduces them to the idea of proportionality and teaches them that experimental laws are only valid provided certain conditions are met. They are introduced to the idea of pressure along with applications that are dependent on the concept. Students find out how to calculate the moment of a force and how to establish the Principle of Moments through practical investigation. They are introduced to the meaning of centre of gravity and learn how it affects the stability of an object.

Hooke's law

When a helical spring is loaded, it stretches.
The natural length is the normal length of the spring without a load on it.
The extended length is the length of the spring when loaded.
The difference between the extended and natural lengths is known as the extension:

extension = extended length − natural length

Prescribed practical

Investigating how the extension of a helical spring depends on the applied force

Aims
* to draw a load–extension graph
* to determine the spring constant

Apparatus
* helical spring
* slotted masses
* metre ruler

Method
1 Prepare a table of results as shown in Table 3.1.
2 Measure the natural length of the spring.
3 Add a 100 g (weight = 1.0 N) mass to hang from the spring.
4 Measure the extended length of the spring.

24

3 APPLICATIONS OF FORCES

Show you can

a) A student is investigating Hooke's law. She applied different loads to the same helical spring. She obtained the following incomplete set of results.

Load/N	0	3	6	9	12
Length of spring/cm	6	8	10	12	14
Extension/cm					

i) Copy the student's table of results and complete the last row.
ii) Explain whether or not Hooke's law was obeyed in this experiment.

Test yourself

1 The table shows the total length of a spring obeying Hooke's law when different loads are applied.

Load/N	Total length/cm
2	12
3	15

a) What extension is produced in the spring by a load of 1 N?
b) Calculate the original length of the spring.
2 The spring in a chest expander has a natural length of 24 cm. A force of 1 N stretches the spring 0.4 cm. Calculate the force needed to stretch the spring to a total length of 60 cm.
3 The following results were obtained from a stretching experiment.

Force on the spring/N	0	1	2	3	4	5	6
Extension/cm	0	1.5	3.0	4.5	6.0	7.5	9.0
Force on the spring/N	0	1	2	3	4	5	6
Extension/cm	0	3.5	7.5	11.5	15.5	18.5	20

Plot graphs of force against extension for these results and mark any regions that follow Hooke's law.

Pressure

Figure 3.3 illustrates a concrete slab lying on each of its sides on soft ground. The weight of the slab is the same no matter which side it is resting on. However the effect on the soft ground depends on the area of contact.

▲ **Figure 3.3** A concrete slab lying on each of its sides

When the slab is lying on side A, the area of contact is very large but the force per metre square is very small, hence the effect on the soft ground will be minimal. In contrast, the force per square metre exerted by side C is very large, so the effect on the soft ground will be large.

We use the term pressure to describe how force is distributed normally (at right angles) over an area. If the force is spread out over a large area, as in the case of side A, we say that the pressure exerted on the ground is small. If the force is concentrated on a small area, as in the case of side C, we say that the pressure is large.

Pressure is defined as the ratio of the normal force to the area of contact.

$$\text{Pressure} = \frac{\text{Force}}{\text{area}} \text{ or } P = \frac{F}{A}$$

26

Tip

In calculations, the units for pressure may be in N/cm² or even N/mm². Use the units that the question provides – you will not be expected to convert mm² or cm² to m².

Example

A lorry has a side with an area of 40 m². The wind exerts a pressure of 500 Pa on the side of the lorry. Calculate the force exerted by the wind on the lorry.

Answer

$p = \frac{F}{A}$

$F = P \times A$

$F = 500 \times 40$

$F = 20\,000\,\text{N}$

Show you can

Explain each of the following:
a) You cannot push your thumb through a wooden desk, but with the same force, you can push a drawing pin into the wood.
b) When a firefighter rescues a dog that has fallen through the ice on a frozen lake, they put their ladder on the ice first and then crawl out to the dog on the ladder.
c) A heavy battle tank will not sink into soft ground.
d) A carpenter will sharpen his chisel before he starts work.

Practicals

These practical tasks contain full instructions on apparatus, method and results analysis and will help develop your practical skills.

CCEA's prescribed practicals are clearly highlighted.

Examples

Examples of questions and calculations that feature full workings and sample answers.

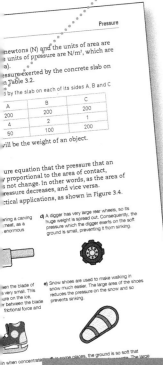

Answers

Answers for all questions in this book can be found online at: www.hoddereducation.co.uk/cceagcsephysics

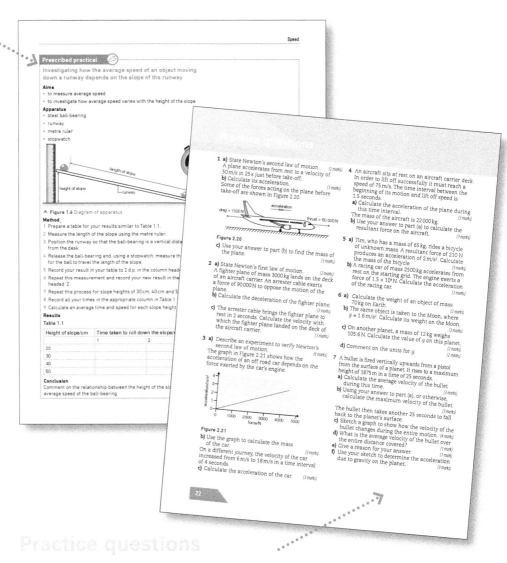

Practice questions

You will find Practice questions at the end of every chapter. These follow the style of the different types of questions you might see in your examination and have marks allocated to each question part.

Level coding

If you are taking GCSE Double Award Foundation-tier you need to study *only* the material with no bars.

If you are taking GCSE Double Award Higher-tier you need to study the material with no bars, plus: the material with the purple H bar and the material with the orange bar.

If you are taking GCSE Physics Foundation-tier you need to study the material with no bars, plus: the material with the light purple F bar and the material with the orange bar.

If you are taking GCSE Physics Higher-tier you need to study all material in the book, including the material marked with the light purple H bar.

1 Motion

Specification points

This chapter covers sections 1.1.1 to 1.1.7 of the GCSE Physics specification and 1.1.1 to 1.1.6 of the Double Award specification. The relationships between distance, average speed, time and rate of change of speed through practical work are covered. It also introduces graphical methods of describing motion. At the higher tier, students are introduced to the concept of vectors and scalars, and they learn about the terms displacement, velocity and acceleration.

Distance and displacement

Distance is the separation between two points. For example, the distance between Belfast and Coleraine is 75 km (see Figure 1.1).

We define displacement as distance in a specified direction. The displacement of Belfast from Coleraine is 75 km south-east. Displacement is represented by an arrow – the length of the arrow is proportional to the distance, and the direction of the arrow is in the same direction as the displacement (see Figure 1.2).

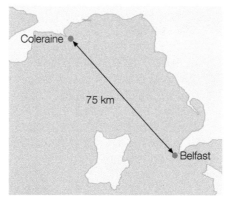

▲ **Figure 1.1** The distance between Coleraine and Belfast

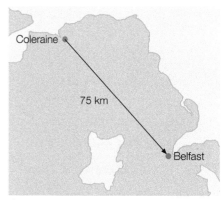

▲ **Figure 1.2** The displacement of Belfast from Coleraine

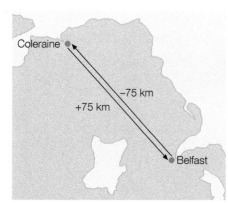

▲ **Figure 1.3** The displacement on a return journey is 0 km

The return journey from Coleraine to Belfast and then back again is a distance of 150 km, but the displacement is 0 km.

The addition of +75 km to –75 km = 0 km (see Figure 1.3).

We say that distance is a scalar quantity – a quantity with size only – whereas displacement is a vector quantity because it has size and direction.

Speed

If a car travels between two points on a road, its speed can be calculated using the formula:

$$\text{speed} = \frac{\text{distance}}{\text{time}}$$

If distance is measured in metres (m) and time in seconds (s), speed is measured in metres per second (m/s).

For example, if a car travels from Coleraine to Belfast in 2 hours, its speed is:

$$\frac{75\,\text{km}}{2\,\text{h}} = 37.7\,\text{km/h}$$

It is unlikely that the car travelled at exactly 37.5 km/h for the whole journey, so this is instead the average speed. The formula for average speed is:

$$\text{average speed} = \frac{\text{total distance}}{\text{time}}$$

To find the actual (or instantaneous) speed at a particular moment in time, we would need to know the distance travelled in a very short interval of time.

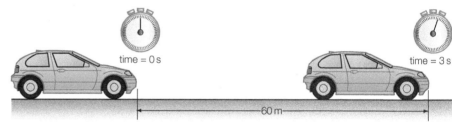

time = 0 s   time = 3 s

60 m

▲ **Figure 1.4** Calculating instantaneous speed at a moment in time

Investigating how the average speed of an object moving down a runway depends on the slope of the runway

Aims
- to measure average speed
- to investigate how average speed varies with the height of the slope

Apparatus
- steel ball-bearing
- runway
- metre ruler
- stopwatch

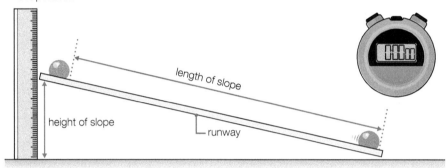

▲ **Figure 1.6** Diagram of apparatus

Method
1 Prepare a table for your results similar to Table 1.1.
2 Measure the length of the slope using the metre ruler.
3 Position the runway so that the ball-bearing is a vertical distance of 20 cm from the desk.
4 Release the ball-bearing and, using a stopwatch, measure the time taken for the ball to travel the length of the slope.
5 Record your result in your table to 2 d.p. in the column headed '1'.
6 Repeat this measurement and record your new result in the column headed '2'.
7 Repeat this process for slope heights of 30 cm, 40 cm and 50 cm.
8 Record all your times in the appropriate column in Table 1.1 to 2 d.p.
9 Calculate an average time and speed for each slope height.

Results

Table 1.1

Height of slope/cm	Time taken to roll down the slope/s			Average speed/cm/s
	1	2	Average	
20				
30				
40				
50				

Conclusion
Comment on the relationship between the height of the slope and the average speed of the ball-bearing.

Example

A car moves from 10 m/s to 30 m/s in a time of 5 seconds. Calculate its rate of change of speed.

Answer

initial speed = 10 m/s
final speed = 30 m/s
time = 5 s

$$\text{rate of change of speed} = \frac{\text{final speed} - \text{initial speed}}{\text{time taken}}$$

$$= \frac{30 - 10}{5}$$

$$= 4 \text{ m/s}^2$$

Show you can (?)

Prove that 72 km/h is equivalent to 20 m/s.

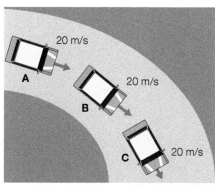

▲ **Figure 1.8** The velocity of the car changes as it goes around the bend

Example

A stone is thrown vertically downwards so that its velocity increases from 2 m/s to 24 m/s. Calculate the average velocity of the stone.

Answer

final velocity = 24 m/s
initial velocity = 2 m/s

$$\text{average velocity} = \frac{\text{initial velocity} + \text{final velocity}}{2}$$

$$= \frac{2 + 24}{2}$$

$$= 13 \text{ m/s}$$

Rate of change of speed

If the speed changes over a period of time (t) it is possible to calculate the rate of change of speed with respect to time, using the formula:

$$\text{rate of change of speed} = \frac{\text{final speed} - \text{initial speed}}{\text{time taken}}$$

This rate of change of speed is a scalar quantity and its units are m/s².

Velocity

Whereas speed is the distance travelled in unit time, velocity is the distance travelled in unit time in a **specified direction**.

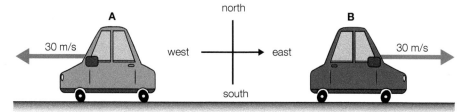

▲ **Figure 1.7** These two cars have the same speed, but different velocities

In Figure 1.7, car A has the same speed as car B, but a different velocity. Car A's velocity is 30 m/s due west, while car B has a velocity of 30 m/s due east.

Speed is a scalar quantity but velocity is a vector quantity.

Since displacement is the distance travelled in a specified direction, we can rewrite the formula for velocity as:

$$\text{average velocity} = \frac{\text{displacement}}{\text{time}}$$

The units for speed and velocity are the same, metres per second (m/s). Occasionally, you will see the units of kilometres per hour (km/h).

The car in Figure 1.8 is moving at a steady, or constant, speed of 20 m/s as it goes around a bend.

The speed of the car at A, B and C is 20 m/s, but the velocity changes as it travels from A to B to C. This is because velocity is a vector quantity, and although the size of the velocity may be constant at 20 m/s, its direction is constantly changing, so its velocity is constantly changing.

Another formula for average velocity is:

$$\text{average velocity} = \frac{\text{initial velocity} + \text{final velocity}}{2}$$

Acceleration

When the velocity of a body increases or decreases, we say it accelerates. Consider the example in Figure 1.9.

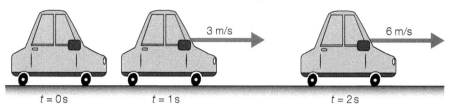

$t = 0s$ $t = 1s$ $t = 2s$

▲ **Figure 1.9** This car is accelerating

The car starts from rest (initial velocity = 0 m/s), but after 1 second its velocity has increased to 3 m/s. After 2 seconds its velocity has increased by 3 m/s to 6 m/s. We say that the car's velocity increases by 3 m/s in 1 second due east – i.e. its acceleration is 3 m/s² due east.

We can define acceleration as the change in velocity in unit time:

$$\text{acceleration} = \frac{\text{final velocity} - \text{initial velocity}}{\text{time}}$$

Acceleration is measured in metres per second squared, written as m/s². Since acceleration is a vector quantity, it can be shown using an arrow of appropriate length and direction.

Most importantly, a '+' or '–' sign can be used to indicate whether the velocity is increasing or decreasing. For example:

+3 m/s² (velocity increasing by 3 m/s every second)

–3 m/s² (velocity decreasing by 3 m/s every second)

A negative acceleration is called a **deceleration**.

A uniform acceleration means a constant (steady) acceleration.

If v = final velocity, u = initial velocity, and t = time taken, then $v - u$ = change in velocity.

So acceleration, a, can be found using $a = \frac{v - u}{t}$, which rearranges to give $v = u + at$.

Show you can

a) A train has an acceleration of 3 m/s². What does this tell you about the velocity of the train?

b) A bus has a deceleration of 2 m/s². What does this tell you about the velocity of the bus?

Example

1 A car starts from rest. After 10 seconds, it is moving with a velocity of 15 m/s. Calculate its acceleration.

Answer

$v = 15$ m/s

$u = 0$ m/s

$t = 10$ s

$a = \dfrac{15 - 0}{10}$

$= 1.5$ m/s²

2 A ball is dropped and accelerates downwards at a rate of 10 m/s² for 5 seconds. By how much will the ball's velocity increase?

Answer

$t = 5$ s

$a = \dfrac{v - u}{t}$

$10 = \dfrac{v - u}{5}$

$v - u = 50$ m/s

Test yourself

4 A car takes 8 s to increase its velocity from 3 m/s to 30 m/s. What is its acceleration?

5 A motorbike, travelling at 25 m/s, takes 5 s to come to a halt. What is its deceleration?

6 An aircraft on take-off has a uniform acceleration of 4 m/s².
 a) What velocity does the aircraft gain in 5 s?
 b) If the aircraft passes a point on the runway at a velocity of 28 m/s, what will its velocity be 8 s later?

7 A ball is thrown vertically upwards in the air, leaving the hand at 30 m/s. The acceleration due to gravity is 10 m/s². Figure 1.10 shows the positions of the ball over time and a data table.

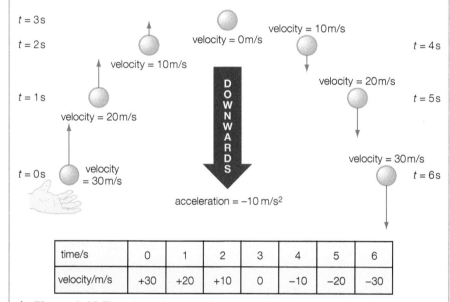

time/s	0	1	2	3	4	5	6
velocity/m/s	+30	+20	+10	0	−10	−20	−30

▲ **Figure 1.10** The changing velocity of a ball thrown into the air

Draw a graph to show the motion of the ball. Plot velocity on the y-axis and time on the x-axis.

Graphs and motion

Graphs are a very useful way of displaying the motion of objects. There are two main types of graphs used:

▶ distance–time graph
▶ speed–time graph.

Distance–time graphs

A **distance–time graph** is a plot of distance on the y-axis versus time on the x-axis.

The simplest type of distance–time graph is shown in Figure 1.11.

This graph illustrates that although the time increases steadily, the distance travelled does not change. The body must be **stationary**.

A horizontal line on a distance–time graph means that the body is stationary.

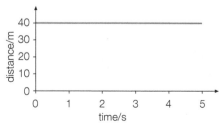

▲ **Figure 1.11** A distance–time graph

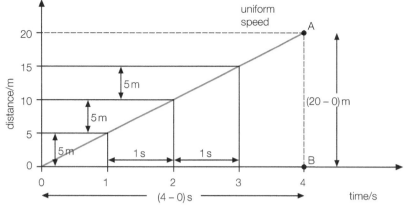

▲ **Figure 1.12** A distance–time graph showing uniform speed

In contrast, the graph in Figure 1.12 shows that the distance is increasing by 5 m every second, i.e. the body is travelling with uniform speed, covering an equal distance in equal units of time.

The slope or gradient of a distance–time graph represents the speed. The speed of the body in Figure 1.12 is 5 m/s.

slope of the graph $= \dfrac{AB}{OB}$

$$= \dfrac{20 - 0}{4 - 0}$$

$$= 5\,\text{m/s}^2$$

When the speed is changing, the slope of the distance–time graph changes. In Figure 1.13, the slope is increasing, which means that the body is accelerating.

Speed–time graphs

A speed–time graph is a plot of speed on the *y*-axis versus time on the *x*-axis.

An example of a speed–time graph for a bicycle is shown in Figure 1.14.

In the first 4 seconds of its motion, the speed of the bicycle increases steadily. The gradient of the line OA is the rate of change of speed.

gradient $= \dfrac{\text{change in } y}{\text{change in } x}$

$$= \dfrac{8 - 0}{4 - 0}$$

$$= 2\,\text{m/s}^2$$

From time $t = 4$ seconds to time $t = 10$ seconds, the speed of the bicycle remains constant (steady) at 8 m/s. There is no rate of change of speed for the bicycle. The area under a speed–time graph represents the distance travelled.

For example, in Figure 1.14 from time $t = 4$ s to time $t = 10$ s:

distance travelled at constant speed = area of rectangle CABD

$$= CA \times CD$$

$$= 8 \times 6$$

$$= 48\,\text{m}$$

The distance travelled by the bicycle during this time is 48 m.

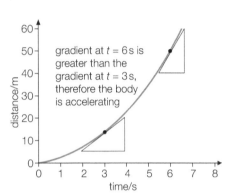

gradient at $t = 6$ s is greater than the gradient at $t = 3$ s, therefore the body is accelerating

▲ **Figure 1.13** A distance–time graph showing acceleration

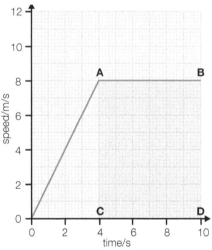

▲ **Figure 1.14** Speed–time graph for a bicycle

Show you can

Calculate the distance travelled in the first 4 seconds of the motion of the bicycle.

Displacement–time graphs

A **displacement–time graph** is a plot of displacement on the y-axis versus time on the x-axis.

This type of graph is very similar to a distance–time graph. The similarity is that a horizontal line on a displacement–time graph means that the body is stationary.

In contrast, the slope or gradient of a displacement–time graph represents velocity and not speed.

Consider Figure 1.15, which represents the movement of a toy car.

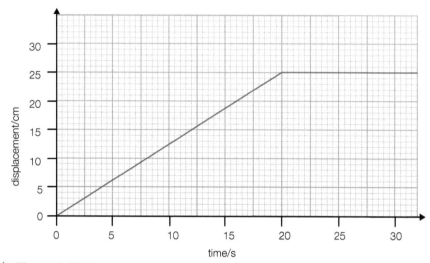

▲ **Figure 1.15** Displacement–time graph for a toy car

From time $t = 0$ s to time $t = 20$ s, the displacement of the toy car increases steadily at a rate of 25 cm in 20 s, which is 1.25 cm/s. The velocity is 1.25 cm/s.

From time $t = 20$ s onwards, the displacement is 25 cm. The toy car is stationary during this time.

Velocity–time graphs

A **velocity–time graph** is a plot of velocity on the y-axis versus time on the x-axis.

The simplest type of velocity–time graph is shown in Figure 1.16.

This graph illustrates that while time is increasing, the velocity remains at a constant (steady) 30 m/s. The car is not accelerating.

In addition, the area under a velocity–time graph represents the displacement travelled.

Tips ↻

As soon as you see a velocity–time graph, two thoughts should spring to mind:
▶ The gradient is the acceleration.
▶ The area between the line and the time axis is the displacement.

For example, in Figure 1.16:

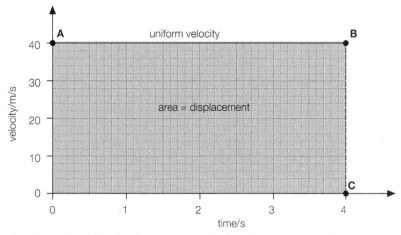

▲ **Figure 1.16** A velocity–time graph showing uniform velocity

area of rectangle OABC = OA × OC

$$= 40 \times 4$$

$$= 160 \text{ m}$$

The displacement travelled by the car is 160 m in a specified direction.

In Figure 1.17, OD is the velocity–time graph for a body accelerating uniformly from rest.

$$\text{acceleration} = \frac{\text{change in } y}{\text{change in } x}$$

$$= \frac{15 - 0}{3 - 0}$$

$$= 5 \text{ m/s}^2$$

The slope or gradient represents the acceleration of the body. Furthermore, the area of the triangle OCD gives the displacement travelled.

displacement = area of triangle OCD

$$= ½ \times OC \times CD$$

$$= ½ \times 3 \times 15$$

$$= 22.5 \text{ m}$$

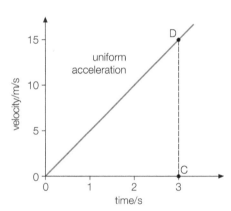

▲ **Figure 1.17** A velocity–time graph showing uniform acceleration

Test yourself ✎

8 Copy and complete this table.

Name of quantity	Definition	Scalar or vector
Distance		scalar
Displacement		
Speed	Rate of change of distance with respect to time	scalar
Velocity		vector
Acceleration	Rate of change of velocity with respect to time	

1 Paul and Jim set off at the same time from their separate houses to travel to a nearby shop. This table shows the distances travelled by Paul to the shop.

Distance travelled by Paul/m	0	3	6	9	12	15	18	21
Time elapsed/s	0	1	2	3	4	5	6	7

a) Draw a graph of distance against time for Paul's journey. *(3 marks)*

The table below shows the distances travelled by Jim to the same shop.

Distance travelled by Jim/m	0	2	4	6	8	10	12	14
Time elapsed/s	0	1	2	3	4	5	6	7

b) Draw a graph of distance against time for Jim's journey on the same axes. *(3 marks)*
c) Use the graphs to answer the following questions.
   i) Which person is going faster? *(1 mark)*
   ii) How long does it take Paul and Jim to travel 11 m? *(1 mark)*
   iii) How far apart are Paul and Jim after 2.5 s? *(2 marks)*
   iv) Is Paul's speed steady? *(1 mark)*
   v) What is Jim's average speed? *(3 marks)*

2 Study the velocity–time graph (Figure 1.18) and describe in words the motion of the object. *(6 marks)*

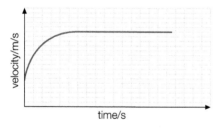

Figure 1.18

3 Figure 1.19 shows a velocity–time graph for a car accelerating away from a junction. Calculate:
a) the acceleration during the first 5 s *(3 marks)*
b) the total displacement. *(4 marks)*

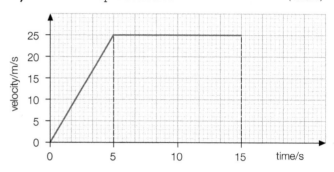

Figure 1.19

4 The graph in Figure 1.20 represents a journey in a lift in a hospital.

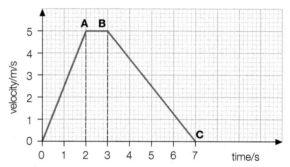

Figure 1.20

a) Briefly describe the motion represented by
   i) OA
   ii) AB
   iii) BC. *(3 marks)*
b) Use the graph to calculate:
   i) the initial acceleration of the lift *(3 marks)*
   ii) the total distance travelled by the lift *(4 marks)*
   iii) the average speed of the lift for the whole journey. *(3 marks)*

Use a graphical method or formula to answer questions 5–11.

5 A car accelerates at 3 m/s² for 10 seconds. It started with a velocity of 20 m/s. Calculate its final velocity. *(3 marks)*

6 An ice-skater moves off from rest (this means that $u = 0$ m/s) with a uniform acceleration of 3.0 m/s². What is her speed and distance travelled after 10 s? *(5 marks)*

7 A stone is thrown vertically upwards with a velocity of 20 m/s. Find how high it will go and the time taken to reach this height (assume $a = -10$ m/s^2). *(6 marks)*

8 A stone is dropped down an empty mine shaft. It takes 3 seconds to reach the bottom. Assuming that the stone falls from rest and accelerates at 10 m/s^2, calculate:
   a) the maximum speed reached by the stone before hitting the bottom *(3 marks)*
   b) the average speed of the stone in flight *(3 marks)*
   c) the depth of the mine shaft. *(3 marks)*

9 A helicopter at a height of 500 m drops a package which falls to the ground in a time of 10 s. Neglecting air resistance and assuming the acceleration is constant and equal to 10 m/s^2, calculate:
   a) the time taken for the package to reach the ground *(3 marks)*
   b) the average velocity of the package. *(3 marks)*

10 A ball is thrown vertically upwards into the air with a velocity of 50 m/s. Neglecting air resistance and assuming the acceleration is constant and equal to -10 m/s^2, calculate:
   a) the time taken to reach maximum height *(3 marks)*
   b) the maximum height reached by the ball. *(3 marks)*

11 A cyclist accelerates from rest at 3 m/s^2.
   a) What is his speed after 5 s? *(3 marks)*
   He then decelerates at 0.5 m/s^2.
   b) How long will it take for his speed to reach zero? *(3 marks)*
   c) Draw a velocity–time graph for this motion. *(3 marks)*

12 John makes a return journey to his local shop to buy a newspaper. In Figure 1.21, Graph A is the distance–time graph and Graph B is the displacement–time graph for the journey.
   a) Copy and complete Graph B. *(1 mark)*
   b) What is the average velocity for John's journey? *(1 mark)*

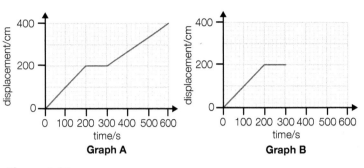

Figure 1.21

13 A sponge ball is allowed to drop from rest and hits the ground after 3.0 seconds. It rebounds to a new height after a further 1.5 seconds. The velocity–time graph of the motion is shown in Figure 1.22.

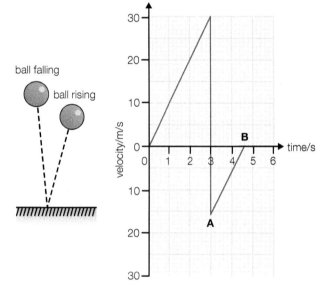

Figure 1.22

The part of the graph AB represents the ball rebounding.
   a) Why is this part of the graph drawn below the time axis? *(1 mark)*
   b) Use the graph to find the height of the rebound. *(3 marks)*
   c) When a ball falls freely under the force of gravity, its acceleration is known as the acceleration of free fall. Use the graph to calculate the acceleration of free fall. *(3 marks)*

Balanced and unbalanced forces

A force has both size and direction. It is another example of a vector quantity. The size of the force is measured in **newtons** (N). When drawing force diagrams, we represent the direction of the force with an arrow and the size of the force by drawing the length of the arrow to scale. By convention, forces acting to the right are positive, and forces acting to the left are negative.

If the forces are equal in size and opposite in direction, then the forces are balanced. Balanced forces do not change the velocity of an object.

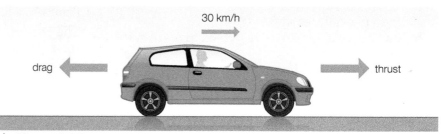

30 km/h

drag

thrust

▲ **Figure 2.1** Forces acting upon a car travelling at a steady speed

Figure 2.1 shows a car travelling at a steady speed of 30 km/h in a straight line under the action of two equal and opposite forces: the thrust exerted by the engine, and air resistance (drag).

If an object is stationary (not moving), it will remain stationary.

In a tug of war like that in Figure 2.2, two teams pull against each other. When both teams pull equally hard, the forces are balanced and the rope does not move. But when one team starts to pull with a larger force, the rope moves. This is how we can tell that the two forces are no longer balanced.

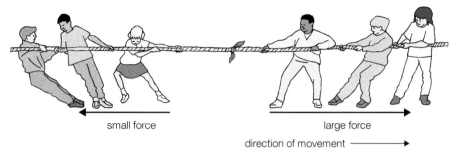

small force   large force

direction of movement ────────➤

▲ **Figure 2.2** The forces in this tug of war are unbalanced as the team on the right is pulling with a larger force

Unbalanced forces will change the velocity of an object. Since velocity involves both speed and direction, unbalanced forces can make an object speed up, slow down or change direction.

Unbalanced forces applied to the handlebars will make the cyclist in Figure 2.3 change direction. This means the velocity of the cyclist will change, even though the speed may stay the same.

An object will only accelerate when an unbalanced force acts on it. It then accelerates in the direction of the unbalanced force. In Figure 2.4, if the driving force on a car is greater than the frictional force, the car will accelerate or speed up.

pushing force   pulling force

▲ **Figure 2.3** Unbalanced forces acting upon the handlebars affect velocity

drag   driving force

▲ **Figure 2.4** This car is accelerating

If the driver then decides to apply the brakes (Figure 2.5), the driving force will be smaller than the braking force, and the car will decelerate (or slow down).

braking force   driving force

▲ **Figure 2.5** This car is decelerating

A car is travelling in a straight line along a motorway.

Table 2.1 shows the situations in which there is an unbalanced force on the car.

Table 2.1

Situation	Unbalanced force acting
The car's speed is increasing.	yes
The car's speed is decreasing.	yes
The car's speed is constant.	no
The car starts going round a bend.	yes

Newton's laws

All that we have said about forces so far is summarised by **Newton's first law**:

In the absence of unbalanced forces, an object will continue to move in a straight line at constant speed (its velocity is constant).

Practical activity

An experiment to investigate Newton's first law

Apparatus
▶ linear air track and blower
▶ glider and interrupt card
▶ two light gates and a data logger

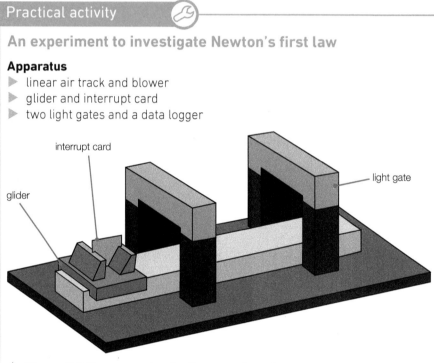

▲ **Figure 2.6** The apparatus for the experiment

Method

1 Set the linear air track on a flat surface and adjust the feet on the air track to make sure that it is level.

2 Measure the length of the interrupt card and enter this in the data logger.

3 Connect up the light gates so that they measure the velocity of the glider at two points.

4 Give the glider a gentle push so that it passes through both light gates.

5 Confirm by looking at the results that the velocity does not change between the two positions of the light gate and so the glider is obeying Newton's first law of motion.

6 Repeat for other velocities and positions of the light gates.

Linking unbalanced forces, mass and acceleration

Look at Figure 2.7. It is possible for one person to push a car, but the acceleration of the car would be small. The more people pushing the car, the larger the acceleration. So, the larger the force, the larger the acceleration.

Now look at Figure 2.8. Even four people would find it difficult to push a van, because the mass of a van is usually far larger than the mass of a car. The larger the mass, the smaller the acceleration.

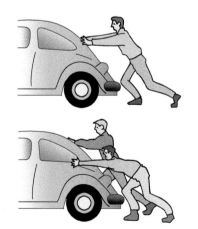

▲ **Figure 2.7** People exerting force upon a car

A

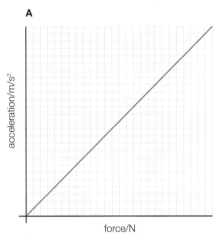

force/N

B

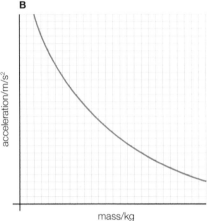

mass/kg

▲ **Figure 2.9** Graphs illustrating Newton's second law

▲ **Figure 2.8** People trying to push a van, a much larger mass

The size of the resultant or unbalanced force needed to accelerate a mass can be worked out using **Newton's second law**.

resultant force (N) = mass (kg) × acceleration (m/s^2)

or

$$F = m \times a$$

This is a simple formula, but it contains a lot of information. You should appreciate that the acceleration is directly proportional to the resultant force, i.e. the larger the resultant force, the larger is the acceleration (for a constant mass).

Furthermore, if the size of the resultant force is constant, then the larger the mass, the smaller the acceleration. We say that the mass and the acceleration are inversely proportional to each other.

The graphs in Figure 2.9 illustrate these two concepts.

Newton's second law states that the acceleration of a body is directly proportional to the force applied to it, and the direction of the acceleration is parallel to the direction of the force.

This law also explains why some very large articulated lorries have long braking distances. When the stopping force is constant, the deceleration is inversely proportional to the mass of the lorry.

Practical activity

Investigating Newton's second law

Experiment 1: Investigating the link between acceleration versus force (at constant mass)

Apparatus

► runway
► trolley
► string
► double interrupt mask with both sections the same width
► light gate and data logger
► pulley masses
► balance

Method

1 Prepare a table for results as shown in Table 2.2.

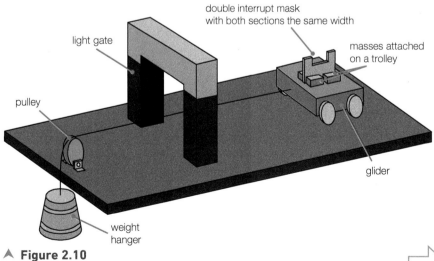

▲ **Figure 2.10**

2 Friction is always present when a trolley rolls along a runway. To compensate for friction, tilt the runway until the trolley moves with a constant speed after it is given a gentle push.

3 Fix the clamped pulley to the end of the bench (see Figure 2.10).

4 Attach a length of string to a dowel rod on the end of the trolley and make a loop in the other end to hang the masses from (the masses falling will provide a constant accelerating force). Pass this over the clamped pulley.

5 Position the light gate in such a way that the mask on top of the trolley passes through it, without hitting anything, before the masses on the end of the string hit the ground

6 Use the light gate to measure the acceleration of the trolley for various driving forces (masses on the mass hanger) from 100 g–500 g, repeating each measurement two times and taking an average. Remember, each 100g mass is equivalent to 1 N.

7 Find the weight of the masses and record this as the value for the resultant force F.

8 Plot a graph of acceleration (y-axis) versus force (x-axis).

Results

Table 2.2

Resultant force F/N	First acceleration/m/s^2	Second acceleration/m/s^2	Average acceleration/m/s^2
1.0			
2.0			
3.0			
4.0			
5.0			
6.0			

Conclusion

▶ The graph is a ....showing that the acceleration is directly proportional to the resultant force.

Is it possible to calculate the mass of the trolley?

Experiment 2: Investigating the link between acceleration and mass (at constant force)

Apparatus

▶ Use the same apparatus as shown in Figure 2.10 but arrange the slotted masses as directed below.

Method

1 Prepare a table for results as shown in Table 2.3.

2 Friction is always present when a trolley rolls along a runway. To compensate for friction adjust the angle of the slope of the runway until the trolley moves with a constant speed after it is given a gentle push.

3 Fix the clamped pulley to the end of the bench.

4 Attach a length of string to the end of the trolley and make a loop in the other end to hang the masses from (the masses falling will provide a constant accelerating force). Pass this over the clamped pulley.

5 Position the light gate in such a way that the mask on top of the trolley passes through it, without hitting the light gate, before the masses on the end of the string hit the ground.

6 Choose a suitable value for the driving force provided by the falling weights, for example 500 g (5 N).

7 Use the light gate to measure the acceleration of the trolley for various masses of trolley by either adding slotted masses to the trolley or stacking trolleys on top of each other.

8 Repeat each measurement three times and take an average.

9 Plot a graph of acceleration (y-axis) versus mass of trolley (x-axis).

Results

Table 2.3

Resultant force F/N	First acceleration/m/s²	Second acceleration/m/s²	Average acceleration/m/s²
1.0			
2.0			
3.0			
4.0			
5.0			
6.0			
7.0			

Conclusion

The graph is non-linear and implies that as the mass increases, the acceleration ....

Example

1 Calculate the force needed to give a train of mass 250 000 kg an acceleration of 0.5 m/s².

Answer

$F = m \times a$

$= 250\,000 \times 0.5$

$= 125\,000\,N$

2 A forward thrust of 400 N exerted by a speedboat enables it to go through the water at constant velocity. The speedboat has a mass of 500 kg. Calculate the thrust required to accelerate the speedboat at 2.5 m/s².

Answer

Note the phrase 'at constant velocity'. This is a clue to using Newton's first law. If the thrust exerted by the engine is 400 N, there must be an equal and opposite force of 400 N due to the drag of the water on the boat. To calculate the force to accelerate the speedboat, we should draw a force diagram (Figure 2.11).

unbalanced force = mass × acceleration

$(F - 400) = 500 \times 2.5$

$F - 400 = 1250$

$F = 1650\,N$

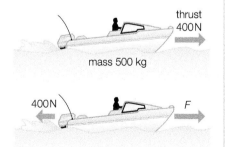

▲ **Figure 2.11** A force diagram

Summary of balanced and unbalanced forces

▶ Balanced forces have no effect on the movement of an object. If it is stationary it will remain stationary; if it is moving it will carry on moving at the same speed and in the same direction.

▶ Unbalanced forces will affect the movement of an object.

▶ An unbalanced force on an object causes its velocity to change (the object accelerates). The greater the force, the greater the acceleration.

▶ The greater the mass of an object, the greater the force needed to make it accelerate.

Test yourself

1 A bicycle and rider have a total mass of 90 kg and travel along a horizontal road at a steady speed. The forward force exerted by the cyclist is 40 N.
   a) Explain why the cyclist does not accelerate.
   b) The rider increases the forward force to 70 N. Calculate the acceleration.

2 A car accelerates at 3.0 m/s² along a road. The mass of the car is 1200 kg and all the resistive forces add up to 400 N. Calculate the forward thrust exerted by the car's engine.

3 Calculate the force of friction on a car of mass 1200 kg if it accelerates at 2 m/s² when the engine force is 3000 N.

4 Figure 2.12 shows the forces on a car of mass 800 kg.
   a) In what direction will the car accelerate?
   b) Calculate the size of the car's acceleration.

5 Look at Figure 2.13. The blades of a helicopter exert an upward force of 25 000 N. The mass of the helicopter is 2000 kg.
   a) Calculate the weight of the helicopter.
   b) Calculate the acceleration of the helicopter.

6 A forward thrust of 300 N exerted by a speedboat engine enables the speedboat to go through the water at a constant speed. The speedboat has a mass of 500 kg. Calculate the thrust required to accelerate the speedboat at 2 m/s².

7 A car and driver are travelling at 24 m/s and the driver decides to brake, bringing the car to rest in 8 seconds. The mass of the car and driver is 1200 kg.
   a) Calculate the deceleration of the car.
   b) Calculate the size of the unbalanced force which brings the car to rest.

8 A cyclist and her bicycle have a combined mass of 60 kg. When she cycles with a forward force of 120 N, she moves at a steady speed. However, when she cycles with a forward force of more than 120 N, she accelerates.
   a) Explain, in terms of forces, why the girl moves at a steady speed when the force is 120 N.
   b) Calculate her acceleration when the forward force is 300 N.

9 A car's brakes are applied and the vehicle's velocity changes from 50 m/s to zero in 5 seconds.
   a) Calculate the acceleration of the car.
   b) The resultant force causing this acceleration is 18 000 N. Calculate the mass of the car.

10 A car of mass 1000 kg is travelling at 20 m/s when it collides with a wall. The front of the car collapses in 0.1 seconds, by which time the car is at rest.
   a) Calculate the deceleration of the car.
   b) Calculate the force exerted by the wall on the car.

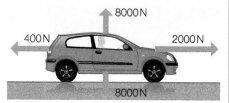

▲ **Figure 2.12** Forces acting upon a car

▲ **Figure 2.13** Forces acting on a helicopter

a) Copy and complete this table.

Table 2.4 Comparing mass and weight

	Mass	Weight
1	is an amount of material	
2		varies from place to place
3	is measured in kg	is measured in ....
4	mass = density × volume	

b) What is the weight of a 70 kg man on
 i) the Earth, where $g = 10$ N/kg?
 ii) the Moon where $g = 1.6$ N/kg?

Mass and weight

In everyday life, the terms 'mass' and 'weight' are used interchangeably. In physics, however, we must be very careful to distinguish clearly between mass and weight.

What is mass?

Mass is defined as the amount of matter in a body. Mass is measured in kilograms (kg). It is a scalar quantity.

What is weight?

Weight is a force and is a measure of the size of the gravitational pull on an object exerted, in our case, by the Earth. Near the surface of the Earth, there is a force of 10 N on each 1 kg of mass. We say that the Earth's gravitational field strength, g, is 10 N/kg.

The weight, W, of an object is the force that gravity exerts on it. The formula for weight is:

weight/N = mass/kg × acceleration due to gravity/m/s²

or, $W = m \times g$

Weight is measured in newtons (N). It is a vector quantity, so it has direction as well as size.

The value of g is roughly the same everywhere on the Earth's surface. But the further you move away from the Earth, the smaller g becomes (see Figure 2.14).

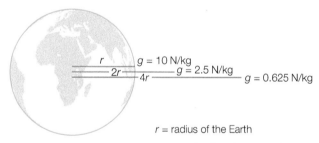

r = radius of the Earth

g = acceleration due to gravity

▲ **Figure 2.14** The gravitational field strength, g, decreases with distance from the Earth

The Moon is smaller than the Earth, and pulls objects towards it less strongly. On the Moon's surface, the value of g is 1.6 N/kg.

In deep space, far away from the planets, there are no gravitational pulls, so g is zero, and therefore everything is weightless.

The size of g also gives the gravitational acceleration, because, from Newton's second law:

$$\text{acceleration} = \frac{\text{force}}{\text{mass}}$$

$$g = \frac{W}{m}$$

So an alternative set of units for g is m/s².

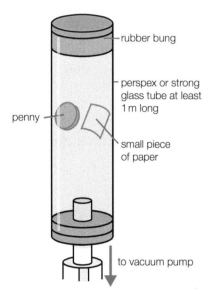

▲ **Figure 2.15** Investigating falling bodies in a vacuum

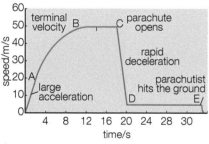

▲ **Figure 2.16** A speed–time graph for a sky-diver

▲ **Figure 2.18** This sky-diver has reached his terminal velocity

Free fall

Galileo is said to have shown that two lead balls of different diameter hit the ground at the same instant when dropped from the top of the leaning tower of Pisa.

In the absence of air resistance, all bodies fall at the same rate of 10 m/s² near the surface of the Earth. It is a common misconception to think that a more massive object falls faster than a less massive one. It is true that there is a greater force on the more massive object, but the acceleration, which is the ratio of force to mass, will be the same for both bodies.

$$a = \frac{F}{m} \text{ or } g = \frac{W}{m}$$

This means that, if there is no air resistance, the speed of a falling object will increase by 10 m/s every second, i.e. its acceleration is 10 m/s². This is known as the acceleration of free-fall, the symbol for which is 'g'. In a vacuum, where there is no air resistance, all falling objects accelerate at the same rate.

When the glass tube in Figure 2.15 is evacuated and then turned upside down, the penny and piece of paper fall together. Although the penny has more mass than the piece of paper, gravity will exert a larger force on the penny, giving both objects the same acceleration, i.e. the ratio of weight to mass is the same for both the penny and the piece of paper.

In the Earth's atmosphere, air resistance does act on a falling body. Air resistance can only be ignored if the force it exerts is very small. When a sky-diver jumps from a plane (Figure 2.17), the forces on them are unbalanced, and so the sky-diver accelerates.

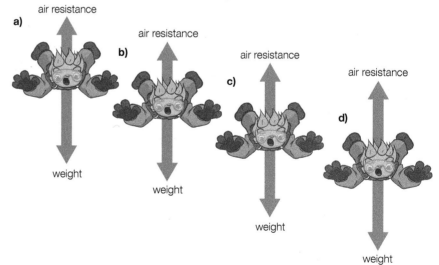

▲ **Figure 2.17** The weight of the sky-diver stays the same, but air resistance increases with speed. Eventually, air resistance = weight, and the sky-diver reaches his terminal velocity

Figure 2.16 shows a speed–time graph for the motion of a sky-diver. The faster the sky-diver falls, the larger the air resistance, so the smaller the acceleration. Eventually the downward force due to gravity and the upward force due to air resistance will be balanced. The sky-diver will stop accelerating and start to fall at a constant speed. At this point, the sky-diver has reached his terminal velocity (Figure 2.18).

a) Julie said, 'My weight is 55 kg.' What is wrong with this statement and what do you think her weight really is?

b) A ball-bearing is gently dropped into a tall cylinder of oil which resists its motion. Describe what will happen to the ball-bearing.

c) An astronaut standing on the surface of the Moon releases a hammer and a feather from the same height. What will happen and why?

d) Why does a parachute slow down a falling parachutist?

e) Explain the shape of each section, AB, BC, CD and DE, of the graph in Figure 2.16.

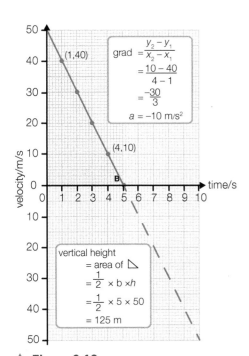

▲ **Figure 2.19**

Opening the parachute increases the air resistance. Since the force of gravity stays the same, it will take less time before the two forces are balanced. This gives the sky-diver a lower terminal velocity (Figure 2.18). Terminal velocity depends on the size of the air resistance force.

Air resistance is affected by the shape of an object. If the sky-diver pulls their arms and legs in line with their body, air resistance is reduced. It will take longer for the upward and downward forces to balance, so the sky-diver reaches a higher terminal velocity. Terminal velocity depends on the shape of an object.

Vertical motion under gravity

When a body is thrown vertically upwards, its motion is opposed by the force of gravity. The velocity of the body will decrease by 10 m/s in each and every subsequent second until its vertical velocity is zero. The body has experienced negative acceleration, more often referred to as **deceleration or retardation**.

 Example

1 A ball is thrown vertically upwards with an initial velocity of 50 m/s. How long will the ball take to reach the top of its motion?

Answer

initial velocity $u = 50$ m/s   final velocity $v = 0$ m/s

time of vertical motion $= t$ acceleration $a = -10$ m/s^2

$$a = \frac{v - u}{t}$$
$$-10 = \frac{0 - 5}{t}$$
$$t = \frac{-50}{-10}$$
$$t = 5\text{ s}$$

This problem could also have been solved graphically by drawing a velocity–time graph for the motion (Figure 2.19).

Time/s	0	1	2	3	4	5	6	7	8	9	10
Velocity/m/s	50	40	30	20	10	0	−10	−20	−30	−40	−50

 Test yourself

11 Use the graph in Figure 2.19 to determine:
   a) the total time the ball is in the air
   b) the velocity with which it strikes the ground
   c) the total distance the ball travelled in the air
   d) the total displacement of the ball.

1 a) State Newton's second law of motion. *(2 marks)*
A plane accelerates from rest to a velocity of 50 m/s in 25 s just before take-off.
b) Calculate its acceleration. *(3 marks)*
Some of the forces acting on the plane before take-off are shown in Figure 2.20.

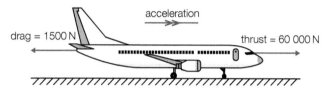

acceleration

drag = 1500 N   thrust = 60 000 N

Figure 2.20

c) Use your answer to part (b) to find the mass of the plane. *(3 marks)*

2 a) State Newton's first law of motion. *(2 marks)*
A fighter plane of mass 3000 kg lands on the deck of an aircraft carrier. An arrester cable exerts a force of 90 000 N to oppose the motion of the plane.
b) Calculate the deceleration of the fighter plane. *(3 marks)*
c) The arrester cable brings the fighter plane to rest in 2 seconds. Calculate the velocity with which the fighter plane landed on the deck of the aircraft carrier. *(3 marks)*

3 a) Describe an experiment to verify Newton's second law of motion. *(6 marks)*
The graph in Figure 2.21 shows how the acceleration of an off road car depends on the force exerted by the car's engine.

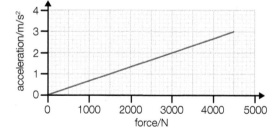

Figure 2.21

b) Use the graph to calculate the mass of the car. *(3 marks)*
On a different journey, the velocity of the car increased from 6 m/s to 18 m/s in a time interval of 4 seconds.
c) Calculate the acceleration of the car. *(3 marks)*

4 An aircraft sits at rest on an aircraft carrier deck. In order to lift off successfully it must reach a speed of 75 m/s. The time interval between the beginning of its motion and lift off speed is 1.5 seconds.
a) Calculate the acceleration of the plane during this time interval. *(3 marks)*
The mass of the aircraft is 22 000 kg.
b) Use your answer to part (a) to calculate the resultant force on the aircraft. *(3 marks)*

5 a) Tim, who has a mass of 65 kg, rides a bicycle of unknown mass. A resultant force of 210 N produces an acceleration of 2 m/s². Calculate the mass of the bicycle. *(3 marks)*
b) A racing car of mass 2500 kg accelerates from rest on the starting grid. The engine exerts a force of 1.5×10^4 N. Calculate the acceleration of the racing car. *(3 marks)*

6 a) Calculate the weight of an object of mass 70 kg on Earth. *(2 marks)*
b) The same object is taken to the Moon, where $g = 1.6$ m/s². Calculate its weight on the Moon. *(3 marks)*
c) On another planet, a mass of 12 kg weighs 105.6 N. Calculate the value of g on this planet. *(2 marks)*
d) Comment on the units for g. *(2 marks)*

7 A bullet is fired vertically upwards from a pistol from the surface of a planet. It rises to a maximum height of 1875 m in a time of 25 seconds.
a) Calculate the average velocity of the bullet during this time. *(3 marks)*
b) Using your answer to part (a), or otherwise, calculate the maximum velocity of the bullet. *(3 marks)*
The bullet then takes another 25 seconds to fall back to the planet's surface.
c) Sketch a graph to show how the velocity of the bullet changes during the entire motion. *(4 marks)*
d) What is the average velocity of the bullet over the entire distance covered? *(1 mark)*
e) Give a reason for your answer. *(1 mark)*
f) Use your sketch to determine the acceleration due to gravity on the planet. *(3 marks)*

8 The boat in Figure 2.22 has a mass of 15 000 kg.

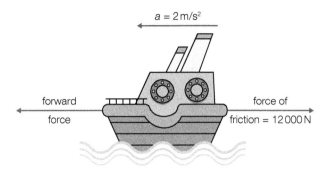

Figure 2.22

The force of friction on the boat is 12 000 N and it is accelerating at 2 m/s². Calculate:
a) the resultant force on the boat *(1 mark)*
b) the forward force from the boat's engines.
(3 marks)

9 An object falls from rest and strikes the ground exactly 1.5 seconds later.
a) At what speed does it hit the ground? *(3 marks)*
b) A ball is thrown vertically upwards with an initial speed of 24 m/s.
i) How long does it take the ball to reach maximum height? *(3 marks)*
ii) What is the maximum height? *(3 marks)*

10 The friction force opposing the motion of a locomotive of mass 25 000 kg is 100 000 N.
a) What forward force must the locomotive provide if it is to travel along a straight, horizontal track at a steady speed of 1.5 m/s²? *(1 mark)*
b) What is the acceleration of the locomotive if the forward force increases to 175 000 N and the friction force is unchanged? *(3 marks)*

3 Applications of forces

Specification points

This chapter covers sections 1.2.13 to 1.2.26 of the GCSE Physics specification (and 1.2.12 to 1.2.23 of the Double Award specification). Students investigate Hooke's law, which introduces them to the idea of proportionality and teaches them that experimental laws are only valid provided certain conditions are met. They are introduced to the idea of pressure along with applications that are dependent on the concept. Students find out how to calculate the moment of a force and how to establish the Principle of Moments through practical investigation. They are introduced to the meaning of centre of gravity and learn how it affects the stability of an object.

Hooke's law

When a helical spring is loaded, it stretches.

The natural length is the normal length of the spring without a load on it.

The extended length is the length of the spring when loaded.

The difference between the extended and natural lengths is known as the extension:

extension = extended length – natural length

Prescribed practical

Investigating how the extension of a helical spring depends on the applied force

Aims

- to draw a load–extension graph
- to determine the spring constant

Apparatus

- helical spring
- slotted masses
- metre ruler

Method

1 Prepare a table of results as shown in Table 3.1.

2 Measure the natural length of the spring.

3 Add a 100 g (weight = 1.0 N) mass to hang from the spring.

4 Measure the extended length of the spring.

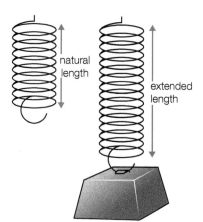

▲ **Figure 3.1** The natural and extended length of a helical spring

5 Calculate and record the extension.

6 Add a second 100 g mass.

7 Repeat measurements and record the results in your table.

Results

Natural length = ... cm

Table 3.1

Load/N	Extended length/cm	Extension/cm
1		
2		
3		
4		
5		
6		

Graphs

Draw a graph of force (N) on the *y*-axis versus extension (cm) on the *x*-axis.

The graph produced should show a straight line such as AB in Figure 3.2.

Conclusion

This experiment shows that the extension of a spring is proportional to the load.

A material that behaves in this way is said to obey **Hooke's law**.

The spring constant = ... N/cm

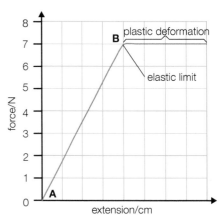

▲ **Figure 3.2** A graph plotting force against extension

Definition of Hooke's law

Extension is proportional to the load, provided the proportional limit is not exceeded.

At point B in Figure 3.2, the spring has reached its **proportional limit**, so no longer obeys Hooke's law. In the region AB of the graph, the spring shows elastic behaviour. This means that when the load is removed, the spring returns to its original length and shape.

If a load of more than 7 N (beyond point B) is applied to this spring, it goes beyond its elastic limit and so changes its shape permanently. When the load is removed, the spring does not return to its original shape. This is called **plastic deformation**.

Formula for Hooke's law

Hooke's law states that the extension of a spring, *e*, is directly proportional to the load, *F*.

Mathematically this is expressed as $F \propto e$ or:

$F = k \times e$

where *k* is the spring constant of the spring.

If *k* is large, the spring is stiff. If *k* is small, the spring is easy to extend.

Show you can

a) A student is investigating Hooke's law. She applied different loads to the same helical spring. She obtained the following incomplete set of results.

Load/N	0	3	6	9	12
Length of spring/cm	6	8	10	12	14
Extension/cm					

i) Copy the student's table of results and complete the last row.
ii) Explain whether or not Hooke's law was obeyed in this experiment.

Test yourself

1 The table shows the total length of a spring obeying Hooke's law when different loads are applied.

Load/N	Total length/cm
2	12
3	15

a) What extension is produced in the spring by a load of 1 N?
b) Calculate the original length of the spring.

2 The spring in a chest expander has a natural length of 24 cm. A force of 1 N stretches the spring 0.4 cm. Calculate the force needed to stretch the spring to a total length of 60 cm.

3 The following results were obtained from a stretching experiment.

Force on the spring/N	0	1	2	3	4	5	6
Extension/cm	0	1.5	3.0	4.5	6.0	7.5	9.0
Force on the spring/N	0	1	2	3	4	5	6
Extension/cm	0	3.5	7.5	11.5	15.5	18.5	20

Plot graphs of force against extension for these results and mark any regions that follow Hooke's law.

Pressure

Figure 3.3 illustrates a concrete slab lying on each of its sides on soft ground. The weight of the slab is the same no matter which side it is resting on. However the effect on the soft ground depends on the area of contact.

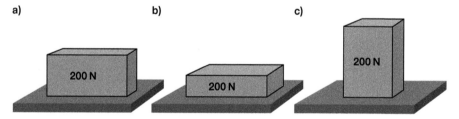

a)   b)   c)

200 N   200 N   200 N

▲ **Figure 3.3** A concrete slab lying on each of its sides

When the slab is lying on side A, the area of contact is very large but the force per metre square is very small, hence the effect on the soft ground will be minimal. In contrast, the force per square metre exerted by side C is very large, so the effect on the soft ground will be large.

We use the term pressure to describe how force is distributed normally (at right angles) over an area. If the force is spread out over a large area, as in the case of side A, we say that the pressure exerted on the ground is small. If the force is concentrated on a small area, as in the case of side C, we say that the pressure is large.

Pressure is defined as the ratio of the normal force to the area of contact.

$$\text{Pressure} = \frac{\text{Force}}{\text{area}} \quad \text{or} \quad P = \frac{F}{A}$$

Tip

In calculations, the units for pressure may be in N/cm² or even N/mm². Use the units that the question provides – you will not be expected to convert mm² or cm² to m².

Example

A lorry has a side with an area of 40 m². The wind exerts a pressure of 500 Pa on the side of the lorry. Calculate the force exerted by the wind on the lorry.

Answer

$p = \dfrac{F}{A}$

$F = P \times A$

$F = 500 \times 40$

$F = 20\,000\,N$

Show you can

Explain each of the following:
a) You cannot push your thumb through a wooden desk, but with the same force, you can push a drawing pin into the wood.
b) When a firefighter rescues a dog that has fallen through the ice on a frozen lake, they put their ladder on the ice first and then crawl out to the dog on the ladder.
c) A heavy battle tank will not sink into soft ground.
d) A carpenter will sharpen his chisel before he starts work.

When the units of force are newtons (N) and the units of area are square metres (m²), then the units of pressure are N/m², which are commonly called **pascals** (Pa).

The calculations for the pressure exerted by the concrete slab on each of its sides are given in Table 3.2.

Table 3.2 The pressure exerted by the slab on each of its sides A, B and C

Side of block	A	B	C
Weight/N	200	200	200
Area on contact/m²	4	2	1
Pressure/Pa	50	100	200

Often, the normal force will be the weight of an object.

$$\text{pressure} = \frac{\text{weight}}{\text{area}}$$

It is clear from the pressure equation that the pressure that an object exerts is inversely proportional to the area of contact, assuming the force does not change. In other words, as the area of contact increases, the pressure decreases, and vice versa.

This fact has many practical applications, as shown in Figure 3.4.

a) A chef will spend time sharpening a carving knife before cutting a joint of meat, as a small area of contact means enormous cutting pressure.

d) A digger has very large rear wheels, so its huge weight is spread out. Consequently, the pressure which the digger exerts on the soft ground is small, preventing it from sinking.

b) The area of contact between the blade of an ice skate and the ice is very small. This results in very large pressure on the ice, producing a layer of water between the blade and the ice, reducing the frictional force and making skating effortless.

e) Snow shoes are used to make walking in snow much easier. The large area of the shoes reduces the pressure on the snow and so prevents sinking.

c) The weight of a woman when concentrated on a stiletto heel results in a very large pressure, so large that floors are easily damaged.

f) In some places, the ground is so soft that houses are built on rafts of concrete. The large area of concrete spreads the weight of the house, so it doesn't sink into the ground.

▲ **Figure 3.4** Practical applications of the relationship between pressure and area of contact

Test yourself

4 A girl weighs 600 N and the total area of her feet is 300 cm². Calculate the pressure she exerts on the floor when she stands on both feet. Give your answer in N/cm².

5 A power washer can produce a fine water jet with a force of 56 000 N over an area of 0.005 m². Calculate the pressure that the washer can exert on the ground.

6 A large metal box is 0.8 m long, 0.5 m wide and 0.4 m deep. The box weighs 320 N. Calculate the areas of each of its faces, and use these to find the maximum and minimum pressures that the box can exert on the ground.

7 A ballet dancer standing on one of her points has a weight of 400 N. Calculate the pressure she exerts on her point if the area of her point is 2 cm².

8 A concrete slab measures 1 m × 0.5 m and exerts a pressure of 1000 Pa on the ground. Calculate the weight of the concrete slab.

9 A man has a weight of 750 N. When standing on one foot, he exerts a pressure of 3 N/cm² on the ground.
a) Calculate the area of contact between his foot and the ground.
b) How will the pressure exerted on the ground by the man be affected if he now stands with both feet on the ground?

10 A tractor has a mass of 3000 kg. The total area of its wheels in contact with the ground is 0.75 m². Calculate the pressure that the tractor exerts on the ground.

Moments

Moment of a force

Door handles are usually placed as far from the hinges as possible so that the door opens and closes easily. A much larger force would be needed if the handle was near the hinges. Similarly, it is easier to tighten or loosen a nut with a long spanner than with a short one.

The **turning effect** or moment of a force depends on two factors:

▶ the size of the force

▶ the distance the force is from the turning point or pivot.

The moment of a force is measured by multiplying the force by the perpendicular distance of the line of action of the force from the pivot (Figure 3.6). This can be written as:

   moment of a force = force × perpendicular distance from the pivot

The unit of the moment of a force is the newton metre (N m), where the force is measured in newtons (N) and the distance from the pivot to the line of action of the force is measured in metres (m).

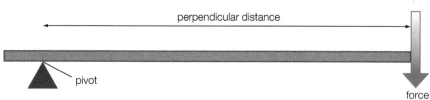

▲ **Figure 3.6** Calculating the moment of a force

Tip

The moment of a force is a vector quantity. The direction of the moment is either clockwise or anticlockwise. Never write 'upwards' or 'downwards'.

Example

Find the moment of a 100 N force applied at a perpendicular distance of 0.3 m from the centre of a nut.

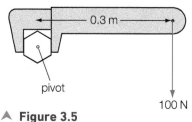

▲ **Figure 3.5**

Answer

turning moment
= force × perpendicular distance

= 100 N × 0.3 m

= 30 N m

The Principle of Moments

The Principle of Moments is as follows:

When a body is in equilibrium, the sum of the clockwise moments about any point equals the sum of the anticlockwise moments about the same point.

The formula that arises from this definition is:

$$F_1 \times d_1 = F_2 \times d_2$$

Another very important consequence of the fact that the body is in equilibrium is that the forces acting on the body in any direction must balance. The upward forces must balance the downward forces. This idea is very useful when solving problems.

Prescribed practical

Investigating the Principle of Moments

Aim
- to measure clockwise and anticlockwise moments

Apparatus
- metre ruler
- slotted masses
- thread
- pivot, such as a string loop attached to clamp and retort stand

Diagram of apparatus

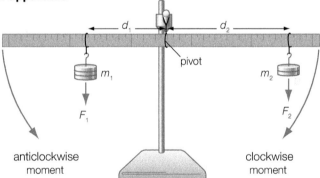

▲ **Figure 3.7** Calculating the moment of a force

Method

1 Suspend and balance a metre ruler at the 50 cm mark using thread.

2 Adjust the position of the thread so that the ruler does not rotate.

3 Hang unequal masses, m_1 and m_2 (100 g slotted masses), from either side of the metre ruler, as illustrated in Figure 3.7.

4 Adjust the position of the masses until the metre rule is balanced (in equilibrium) once again.

5 Gravity exerts forces F_1 and F_2 on the masses m_1 and m_2. Remember that a 100 g slotted mass is equivalent to a weight of 1 N.

6 Record the results in a table such as Table 3.3, and repeat for other loads and distances.

A boy weighing 600 N sits 1.0 m away from the pivot of a see-saw, as shown below.

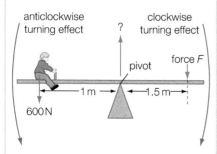

▲ **Figure 3.8** The forces involved in a boy sitting on a see-saw

1 What force 1.5 m from the pivot is needed to balance the see-saw?

Answer

The force F exerts a clockwise turning effect about the pivot, while the boy's weight exerts an anticlockwise turning effect. Since the see-saw is balanced, we can write

$$\frac{\text{clockwise}}{\text{moment}} = \frac{\text{anticlockwise}}{\text{moment}}$$

$$F_1 \times d_1 = F_2 \times d_2$$
$$F_1 \times 1.5\,\text{m} = 600\,\text{N} \times 1.0\,\text{m}$$
$$F_1 = (600 \times 1) \div 1.5$$
$$F_1 = 400\,\text{N}$$

2 Find the size of the upward force exerted by the pivot.

Answer

Since the body is balanced (in equilibrium): the upward force at the pivot = the sum of the downward forces acting on the see-saw

$$= 400\,\text{N} + 600\,\text{N}$$
$$= 1000\,\text{N}$$

7 The force F_1 is trying to turn the metre stick anticlockwise, and $F_1 \times d_1$ is its moment. F_2 is trying to turn the metre stick clockwise, its moment is $F_2 \times d_2$.

8 When the metre stick is balanced (i.e. in equilibrium), the results should show that the anticlockwise moment $F_1 \times d_1$ equals the clockwise moment $F_2 \times d_2$.

Results

Table 3.3

Anticlockwise				Clockwise			
m_1/g	F_1/N	d_1/cm	$F_1 \times d_1$/N cm	m_2/g	F_2/N	d_2/cm	$F_2 \times d_2$/N cm

Conclusion

What do you deduce from the values in columns four and eight of your results?

11 Figure 3.9 shows a car park barrier. The weight of the barrier is 150 N, and its centre of mass is 0.9 m from the pivot.

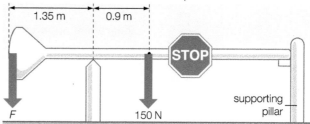

▲ **Figure 3.9** The forces acting upon a car park barrier

a) Calculate the size of the clockwise moment produced by the barrier's weight about the pivot.

b) Calculate the size of the force, F, on the left of the pivot which will just lift the barrier off the supporting pillar.

12 Figure 3.10 shows a uniform metre ruler pivoted at its midpoint. A load of 4 N acts on the right-hand side at a distance of 36 cm from the pivot.

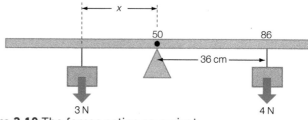

▲ **Figure 3.10** The forces acting on a pivot

Calculate the distance from the pivot where you would place a 3 N weight to balance the metre ruler.

13 Figure 3.11 shows a plan view of a gate pivoted at C. The boy at A is pushing on the gate with a force of 100 N and a man at B is pushing in the opposite direction so that the gate does not move.

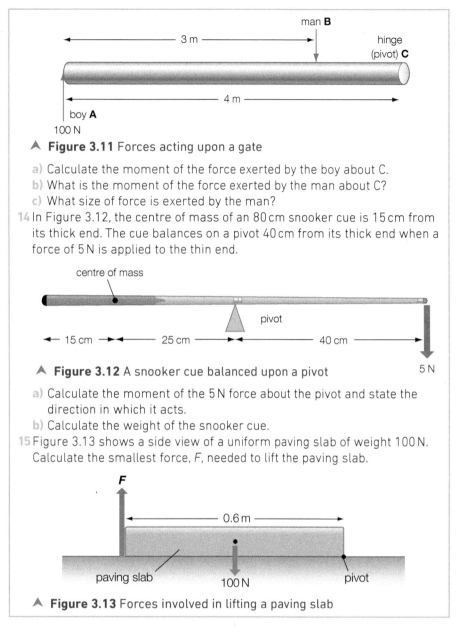

▲ **Figure 3.11** Forces acting upon a gate

a) Calculate the moment of the force exerted by the boy about C.
b) What is the moment of the force exerted by the man about C?
c) What size of force is exerted by the man?

14 In Figure 3.12, the centre of mass of an 80 cm snooker cue is 15 cm from its thick end. The cue balances on a pivot 40 cm from its thick end when a force of 5 N is applied to the thin end.

▲ **Figure 3.12** A snooker cue balanced upon a pivot

a) Calculate the moment of the 5 N force about the pivot and state the direction in which it acts.
b) Calculate the weight of the snooker cue.

15 Figure 3.13 shows a side view of a uniform paving slab of weight 100 N. Calculate the smallest force, *F*, needed to lift the paving slab.

▲ **Figure 3.13** Forces involved in lifting a paving slab

Centre of gravity

All objects have a point at which we can consider all their weight to be concentrated. This point is referred to as the centre of gravity, sometimes called the **centre of mass** of the object.

The centre of gravity is a point through which the whole weight of the body appears to act.

Figure 3.14 shows a metre ruler that is balanced about its midpoint. You could imagine the metre ruler as consisting of a series of 10 cm sections. The mass of each section is pulled towards the centre of the Earth by the force of gravity, so there are several small forces acting on the metre rule. But it is possible to replace all of these forces by a single resultant force acting through the centre of gravity, G. This force may be balanced by the reaction exerted by the pivot, as illustrated in Figure 3.14b.

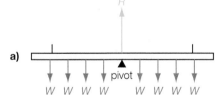

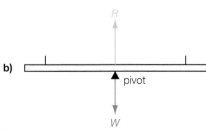

▲ **Figure 3.14** Forces acting on a metre ruler

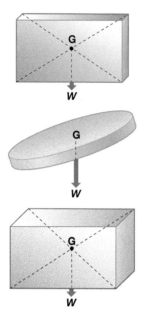

▲ **Figure 3.15** Working out the centre of gravity for regular objects

The centre of gravity of a body may be regarded as the point of balance. If a body has a regular shape, such as a flat disc or a rectangular sheet of metal, then the centre of mass is at its geometrical centre (Figure 3.15).

Flat triangular shapes are a little more difficult. In such cases, lines called **medians** are drawn from the corners of the triangle to the midpoints of the opposite sides. Where the medians intersect is the centre of mass (Figure 3.16).

Finding the centre of gravity of an irregularly shaped lamina

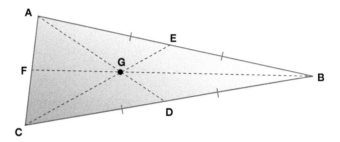

▲ **Figure 3.16** The point where the medians intersect is the triangle's centre of gravity

A lamina is a body in the form of a flat thin sheet.

Figure 3.17 shows a method for finding the centre of mass of a lamina. It is important to realise that when a body is suspended so that it can swing freely, it will come to rest with its centre of gravity vertically below the point of suspension.

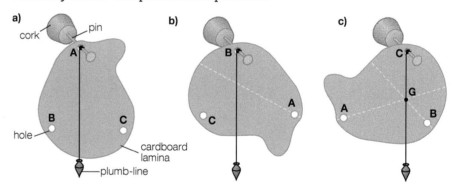

▲ **Figure 3.17** Finding the centre of mass of an irregular lamina

The stages involved in this method are as follows:

1 Hang an irregularly shaped sheet of cardboard from a pin, embedded into a cork.
2 Hang a plumb-line from the same pin.
3 When the cardboard settles, mark the vertical line with a pencil.
4 Repeat from two further points.

The intersection of the vertical lines from the three points of suspension will be the centre of gravity.

Equilibrium and stability

A body is in equilibrium when both the resultant force and resultant turning effect on it are zero. There are three types of equilibrium, which are determined by what happens to the object when it is given a small push.

1 A ball on a flat piece of ground (Figure 3.18) is in neutral equilibrium. When given a gentle push, the ball rolls, keeping its centre of gravity at the same height above its point of contact with the ground.

▲ **Figure 3.18** This ball is in neutral equilibrium with the ground

2 A tall radio mast is in unstable equilibrium (Figure 3.19). It is balanced with its centre of gravity above its base, but a small push from the wind will move its centre of gravity downwards. To prevent the mast toppling, it is stabilised with cables.

3 A car on the road is in stable equilibrium (Figure 3.20a). If the car is tilted (b) the centre of gravity is lifted. In this position, the action of the weight keeps the car on the road. In (c), the centre of gravity lies above the wheels, so the car is in a position of unstable equilibrium. If the car tips further (d) the weight provides the turning effect to turn the car over. Cars with a low centre of gravity and a wide wheelbase are the most stable on the road (Figure 3.21).

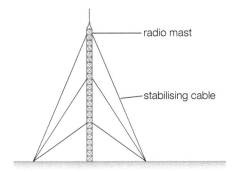

▲ **Figure 3.19** This radio mast is in unstable equilibrium

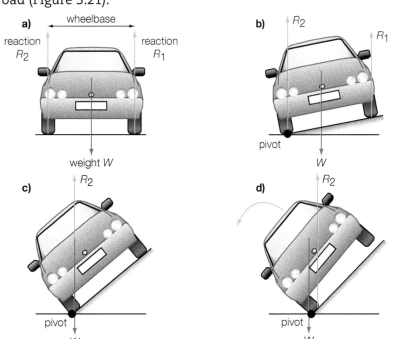

▲ **Figure 3.20** As the car tilts further, it becomes more and more unstable until at position (d), it topples over

▲ **Figure 3.21** This racing car is extremely stable because of its low centre of gravity and wide wheelbase

16 Figure 3.22 shows shapes made from thin sheets of plastic.

⋏ **Figure 3.22** 'Shapes made of thin plastic

a) Copy out the first two shapes and draw construction lines to show where the centre of gravity of each shape is located.

b) In the third shape, the central circular portion (white) has been cut out. If the centre of the circle is at the centre of the square, where will the centre of gravity of this plastic sheet be?

17 a) What is meant by the centre of gravity of an object?

b) Figure 3.23 shows a pencil with a plotting compass attached balancing on its point.
   i) Explain why this happens.
   ii) What would happen if the plotting compass were closed slightly?

c) Figure 3.24 shows a piece of cardboard. Copy the diagram exactly and mark a possible position for the centre of gravity.

18 Figure 3.25 shows a racing car.

a) Copy the diagram and mark with a cross the approximate position of the car's centre of gravity.

b) What two features of the car give it great stability?

19 Figure 3.26 shows cross-sections through two drinking glasses.

a) Copy the diagrams and mark with a cross the approximate position of the centre of gravity of each.

b) Which glass is likely to be more stable?

c) Give two reasons for your answer to part b).

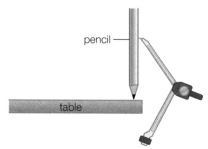

⋏ **Figure 3.23** A pencil balancing on its point using a compass

⋏ **Figure 3.24** A piece of cardboard

⋏ **Figure 3.25** A racing car

⋏ **Figure 3.26** A wine glass and a whisky glass

1 **a)** State Hooke's law. *(3 marks)*
  b) The manufacturers of car seat belts are required by law to test how they behave when different forces are applied to them.
    A particular seatbelt gave the following results.

Load/kN	Seatbelt length/cm	Extension/cm
0	120.5	0.5
0.5	122.0	1.5
1.0	123.5	3.0
1.5	125.0	4.5
2.0	126.5	6.0
2.5	129.0	8.5
3.0	132.5	12.0

  i) What is the natural (unstretched) length of this seatbelt? *(2 marks)*
  ii) Up to what load (in kN) does this seatbelt obey Hooke's law? Explain the reason(s) for your answer. *(3 marks)*
  iii) When the load is removed, the seatbelt always gets shorter. In one case, a load of 5 kN is applied and then removed, and in another case, a load of 1 kN is applied and then removed. How (if at all) does the length of the seatbelt change after the load is removed, in each of these two cases? *(2 marks)*
  iv) When a car is involved in a major accident, it is wise to replace the seat belts. Use your knowledge of how materials behave when stretched to suggest why. *(1 mark)*

2 In an experiment with a helical spring, the following results were recorded.

Load/N	0.0	0.5	1.0	1.5	2.0
Extension/cm	0.0	0.4	0.8	1.2	1.6

  a) Draw a graph of load against extension. *(2 marks)*
  b) Use the graph to find the weight of a metal object that caused an extension of 1.0 cm *(1 mark)*
  c) Calculate the spring constant. *(3 marks)*
  d) Another identical helical spring is connected in series (end to end) with the first spring. Draw a line on the graph to represent the results of the combination of these two springs. *(2 marks)*

3 The table shows the extensions of a spiral spring of natural length 50 mm when increasing loads are attached to it.

Load/N	0	5	10	15	20	25	30	35
Extension/cm	0	2	4	6	8	11	17	26

  a) Draw a graph of extension versus load. *(2 marks)*
  b) Up to what load, according to your graph, does Hooke's law appear to apply? *(1 mark)*
  c) What load should produce an extended length of 25 mm? *(2 marks)*

4) **a)** Explain what is meant by pressure. *(1 mark)*
    An oil jet is used to cut brittle candy into bars.
  b) The jet has a radius of 0.08 mm at the surface of the candy. Calculate the surface area of the candy in contact with the oil jet, giving your answer in mm² and in m².
    Show clearly how you get your answer. *(4 marks)*

    The pressure of the oil jet on the candy is 180 MPa.
  c) What pressure, in pascals, is exerted by the oil jet on the candy? *(1 mark)*
  d) Use your answers to parts **b)** and **c)** to calculate the force which the oil jet exerts on the candy. *(3 marks)*

5 A lorry trailer is 15 m long by 2 m high, as shown in Figure 3.27 below. The force of the wind on the trailer is 150 000 N.

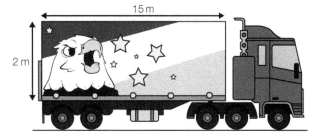

Figure 3.27
  a) Calculate the area of the side of the trailer. *(1 mark)*
  b) Calculate the pressure on the side of the trailer. *(3 marks)*
  c) Explain why similar lorries should avoid high bridges on windy days. *(2 marks)*

6 A wheelbarrow and its load together weigh 600 N (Figure 3.28). The distance between the pivot and the wheelbarrow's centre of mass is 75 cm.

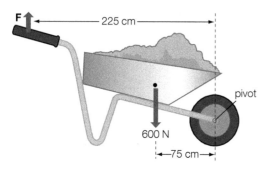

225 cm

F

600 N

pivot

75 cm

Figure 3.28

The distance between the handles and the pivot is 225 cm. *(3 marks)*
Calculate the size of the smallest force, F, needed to lift the wheelbarrow at the handles.

7 a) Explain what is meant by the centre of gravity of an object. *(1 mark)*
Figure 3.29 below shows a wheelbarrow at rest on level ground. The weight of the wheelbarrow and its contents is 1500 N.
b) Use the values on the diagram to calculate the moment of the 1500 N force about the pivot. Show clearly how you get your answer. *(3 marks)*

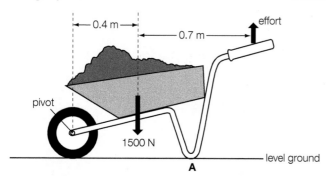

0.4 m

0.7 m

effort

pivot

1500 N

level ground

A

Figure 3.29

c) Use your answer to part **b)** to calculate the effort that must be applied to the handles to lift the wheelbarrow slightly off the ground at A. Show clearly how you get your answer. *(3 marks)*
d) What is the upward vertical reaction (supporting force) from the ground through the pivot when the wheelbarrow is just lifted off the ground by the effort? Show clearly how you get your answer. *(2 marks)*

8 a) State the Principle of Moments. *(2 marks)*
b) A non-uniform plank of wood of length 80 cm is balanced on a pivot as shown in Figure 3.30a.

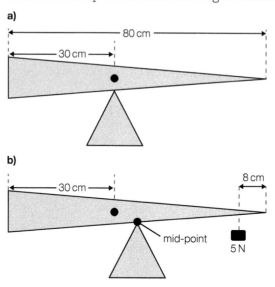

a)

80 cm

30 cm

b)

8 cm

30 cm

mid-point

5 N

Figure 3.30

This demonstrates that the centre of gravity of the plank is 30 cm from the left-hand end, as indicated by a dot. The plank is now moved and rebalanced at its mid-point using a 5 N weight placed 8 cm from the right-hand end, as shown in Figure 3.30b.
i) Calculate the weight of the plank. *(3 marks)*
ii) Calculate the upward force which is now exerted by the triangular support. *(1 mark)*

9 a) A guillotine is used to cut sheets of paper. A constant downward force of 20 N is exerted on the handle (Figure 3.31).

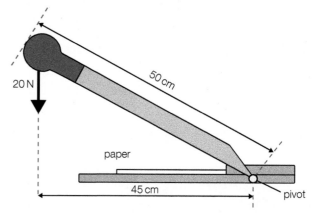

50 cm

20 N

paper

45 cm

pivot

Figure 3.31

Calculate the moment of the 20 N force about the pivot. *(3 marks)*
b) A teapot is placed on a tray and the tray is set on a shelf as shown in Figure 3.32. The tray has a weight of 10 N.

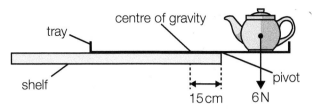

Figure 3.32

The centre of gravity of the tray is 15 cm from the edge of the shelf.
i) Use an arrow to show the direction of the weight of the tray. *(1 mark)*
The teapot weighs 6 N.
ii) Use the principle of moments to find the greatest distance the teapot can be placed from the edge of the shelf without toppling the tray. *(4 marks)*

10 a) Figure 3.33 shows a T-shaped lamina, in which QR is twice as long as AB.

Figure 3.33

Copy the diagram.
i) Mark the centre of gravity of the rectangle ABCD and label it X. *(1 mark)*
ii) Mark centre of gravity of the rectangle PQRS and label it Y. *(1 mark)*
iii) Mark the approximate position of the centre of gravity of the whole shape and label it Z. *(1 mark)*
b) Sketch the shape of a lamina in which the centre of gravity falls outside the shape itself. Mark on the sketch approximately where the centre of gravity lies. *(3 marks)*

11 a) Figure 3.34 shows a solid cone in stable equilibrium.

Figure 3.34

Draw two further diagrams to illustrate a solid cone in:
i) unstable equilibrium
ii) neutral equilibrium. *(2 marks)*

b) Figure 3.35 shows the cross-sections of two similarly shaped table lamps, A and B. The bases in each case are solid.

Figure 3.35

i) Copy the diagrams and mark where you might expect the centre of gravity to be. *(2 marks)*
ii) Which lamp is likely to be more stable? *(1 mark)*
iii) Give two reasons for your answer to part **(ii)**. *(2 marks)*

12 Figure 3.36 shows a bus in two positions. The centre of gravity of the bus is marked G.

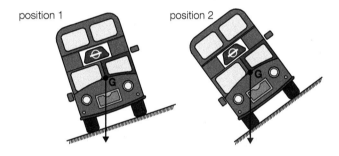

Figure 3.36

a) For each position, describe and explain what happens to the bus. *(4 marks)*
Figure 3.37 shows a long pole being used as a lever to raise a heavy stone block.

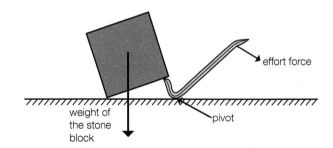

Figure 3.37

The stone block weighs much more than the force the man uses to raise it.
b) Explain carefully how the lever allows him to raise the stone block. *(6 marks)*

13 a) Describe, in detail, an experiment to verify the Principle of Moments. In your description you should include:
- the apparatus used
- how the apparatus is used
- the formula you would use to test the Principle of Moments. *(6 marks)*

Wheel-braces are used to remove wheel nuts (Figure 3.38).

b) i) Explain why wheel-braces are designed so that they may be extended. *(1 mark)*

Figure 3.38

ii) Calculate the moment of a force of 20 N when applied by a wheel-brace of length 0.40 m. *(3 marks)*

4 Density

Specification points

This chapter covers sections 1.3.1 to 1.3.8 in the GCSE Physics specification (and 1.3.1 to 1.3.5 in the Double Award specification). You will investigate the relationship between the volume of a material and its mass leading to the concept of density. You will be introduced to simple kinetic theory which is used to explain the differences between the densities of solids, liquids and gases.

Density

The spectators at a football match are densely packed on the terraces, whereas the footballers on the pitch are well spread out. In a similar way, different materials have different densities. Some materials, such as lead, have large atoms which are very tightly packed together. We say that lead is a very dense material. In contrast, polystyrene has very small, well-spaced-out atoms.

In physics, we compare materials such as lead and polystyrene using the concept of density.

Density is defined as the mass of unit volume of a substance. It is calculated using the formula:

$$\text{density} = \frac{\text{mass of object made of the substance}}{\text{volume of the object}} \text{ or } \rho = \frac{m}{V}$$

The symbol for density is the Greek symbol ρ, which is pronounced 'rho'.

Density is measured in kilograms per cubic metre (kg/m^3) or in grams per cubic centimetre (g/cm^3).

The density of lead is $11\,g/cm^3$, which means that a piece of lead of volume $1\,cm^3$ has a mass of $11\,g$. Therefore, $5\,cm^3$ of lead has a mass of $55\,g$.

If you know the density of a substance, the mass of any volume of that substance can be calculated. This enables engineers to work out the mass (and hence the weight) of a structure if the plans show the volumes of the materials to be used and their densities.

Show you can

Prove that $1\,g/cm^3 = 1000\,kg/m^3$.

Table 4.1 The densities of some common substances

Substance	Density/g/cm³	Density/kg/m³
Aluminium	2.7	2700
Iron	8.9	8900
Gold	19.3	19 300
Pure water	1.0	1000
Ice	0.9	900
Petrol	0.8	800
Mercury	13.6	13 600
Air	0.001 23	1.23

Example

Taking the density of mercury as $14\,g/cm^3$, find:

a) the mass of $7\,cm^3$ of mercury and

b) the volume of $42\,g$ of mercury.

Answer

a) $\rho = \dfrac{m}{V}$

$14 = \dfrac{m}{7}$

$m = 14 \times 7$

$= 98\,g$

b) $\rho = \dfrac{m}{V}$

$14 = \dfrac{42}{m}$

$v = \dfrac{42}{14}$

$= 3\,cm^3$

Measuring density

To determine the density of a substance, we need to measure a) its mass and b) its volume. The density, ρ, will then be given by the ratio of its mass (m) to its volume (V), i.e.

$$\rho = \frac{m}{V}$$

Prescribed practical

Investigating the relationship between the mass and volume of liquids and regular solids

i) Liquids

Aims
- to find the mass of a liquid
- to find the volume of liquid

Apparatus
- $100\,cm^3$ measuring cylinder
- digital balance
- $250\,cm^3$ of water

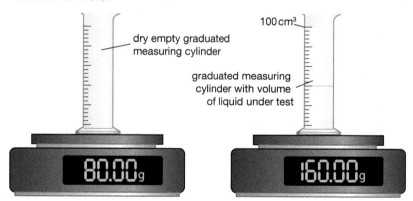

▲ **Figure 4.1** Calculating the density of water

Method

1 Prepare a table for results similar to that shown in Table 4.2.

2 Measure the mass of a $100\,cm^3$ graduated cylinder, using a digital balance.

3 Zero the balance.

4 Pour $20\,cm^3$ of water into the measuring cylinder.

5 Record its mass and volume in the table.

6 Record the mass of 40 cm³, 60 cm³, 80 cm³ and 100 cm³ of water in the same way.

7 The density of the water is found by dividing the mass of the water by the volume of the water.

Results

Table 4.2

Load/N	20	40	60	80	100
Mass/g					
Density/g/cm³					

Conclusion

The average value of density of water is ....

ii) Regularly shaped objects

Aims

- to find the volumes of regularly shaped cubes
- to find the masses of cubes

Apparatus

- digital balance
- ruler
- cubes of material

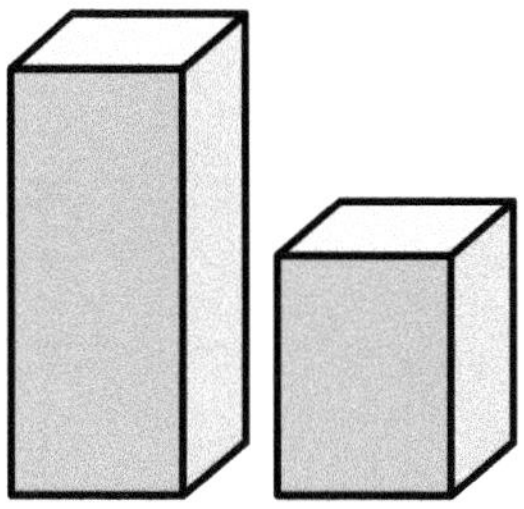

▲ **Figure 4.2** Regularly shaped cubes

Method

1 Prepare a table for results similar to that shown in Table 4.3.

2 Measure and record the dimensions of each cube.

3 Calculate its volume.

4 Measure the mass of each cube using a digital balance.

5 Calculate the density by dividing the mass of the cube by its volume.

Results

Table 4.3

	Block A	Block B	Block C
Length/cm			
Breadth/cm			
Height/cm			
Volume/cm³			
Mass/g			
Density/g/cm³			

Conclusion

The average density of the material is ....

The material is ....

Different volumes of the same material have the same density.

Find the density of irregularly shaped objects

If the shape of the object is too irregular for the volume to be determined using formulae, then a displacement method is used to measure the volume of the irregular solid, as shown in Figure 4.3.

1 The mass of the object is found using a top-pan balance.
2 The volume of the object is equal to the volume of water displaced.
3 The density can be calculated using the formula: $\rho = \dfrac{m}{V}$

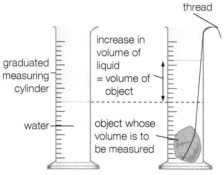

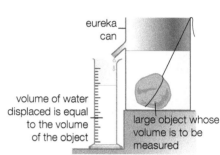

▲ **Figure 4.3a** The volume of a small object can be measured in a measuring cylinder

▲ **Figure 4.3b** Measuring the volume of a large object requires a eureka can

Graphical treatment of density

A graph of mass against volume of a uniform material is always a straight line through the origin. The gradient of the line is equal to the density of a particular substance. The ratio of the co-ordinates of any point on the straight line is the density of that substance (see Figure 4.4).

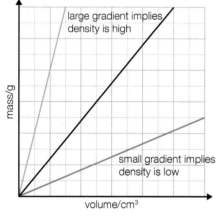

▲ **Figure 4.4** Graph of mass against volume

Test yourself

1 An object with a volume of $3\,cm^3$ weighs $57.9\,g$. Use Table 4.1 on page 40 to find a material that this object could be made from.
2 Aluminium has a density of $2.7\,g/cm^3$.
   a) What is the mass of $20\,cm^3$ of aluminium?
   b) What is the volume of $54\,g$ of aluminium?
3 A piece of steel of mass $120\,g$ has a volume of $15\,cm^3$. Calculate its density.
4 Air has a density of $1.26\,kg/m^3$. Calculate the mass of air in a room of dimensions $10\,m$ by $5\,m$ by $3\,m$.
5 A stone of mass $60\,g$ is lowered into a measuring cylinder, causing the liquid level to rise from $15\,cm^3$ to $35\,cm^3$. Calculate the density of the stone in g/cm^3.
6 The capacity of a petrol tank in a car is $0.08\,m^3$. Calculate the mass of petrol in a full tank if the density of petrol is $800\,kg/m^3$.
7 The mass of an evacuated $1000\,cm^3$ steel container is $350\,g$. The mass of the steel container when full of air is $351.2\,g$. Calculate the density of air.

Explaining the variation in density of solids, liquids and gases using kinetic theory

There are three states of matter – **solids**, **liquids** and **gases**. According to the kinetic theory, matter is made up of very large numbers of atoms and molecules in constant motion.

Look at Figure 4.5. In solids, molecules are packed very closely together. They vibrate about fixed positions and have strong forces of attraction between them. As a result, solids have a fixed shape and volume, and will have a high density.

In liquids, the molecules are close together, but not as close as they are in solids. They can move around in any direction and are not fixed in position. The forces of attraction between them are still quite strong but not as strong as in solids. Liquids have a medium density. This explains why liquids have a fixed volume but take on the shape of the container.

In gases, the molecules are very far apart with large distances between them. They move around very quickly in all directions, and the forces of attraction between them are very weak. Gases have a low density and always completely fill their container.

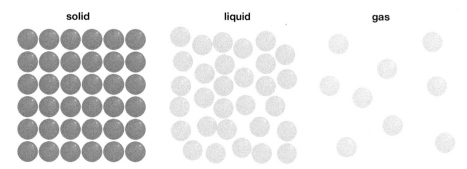

▲ **Figure 4.5** The arrangement of molecules in solids, liquids and gases

Show you can (?)

Copy and complete the table below to summarise the explanation of density using the kinetic theory.

Solids	Liquids	Gases
Molecules vibrate about fixed positions. Molecules have strong .... Molecules are packed very close together, so: solids have a .... density.	Molecules .... Forces of attraction between molecules are still quite strong, but not as strong as in solids. Molecules .... together but not as close as in solids, so: liquids have a medium density.	Molecules are very, very far apart. There are .... forces of attraction between molecules, so: gases have a .... density.

1 a) Explain what is meant by density. *(2 marks)*
b) Describe how you could use a measuring cylinder half-filled with water to find the volume of a bracelet. In your description, state what measurements you would make and what calculation you would carry out. *(4 marks)*
c) A necklace has a volume of 2.4 cm³ and a mass of 46 g. Calculate its density. *(3 marks)*
d) The necklace is made from a metal which is almost 100% pure. Use your answer to part c) and the table below to find out what the metal is.

Metal	Copper	Gold	Lead	Platinum
Density in g/cm³	8.9	19.3	11.3	21.5

(1 mark)

2 Data relating to a particular concrete slab is shown in Figure 4.6.

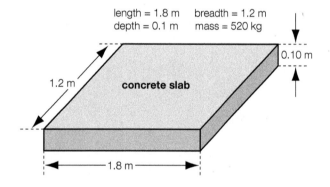

length = 1.8 m   breadth = 1.2 m
depth = 0.1 m   mass = 520 kg

0.10 m

1.2 m   concrete slab

1.8 m

Figure 4.6

a) Use the data to calculate the volume of this concrete slab. *(4 marks)*
b) For bridge construction, the concrete slabs must have a density of at least 2350 kg/m³. Is this particular slab dense enough to be used for bridge construction? *(2 marks)*

3 A brass ingot is 0.6 m wide, 0.5 m tall and 0.2 m long.
a) Find the volume of the brass ingot. *(2 marks)*
Brass has a density of 8400 kg/m³.
b) Calculate the mass of the ingot. *(3 marks)*

4 a) 1 g of water has a volume of 1 cm³. There are 1 000 000 cm³ in 1 m³ of water.
i) What is the mass, in g, of 1 m³ of water? *(1 mark)*
ii) What is the mass, in kg, of 1 m³ of water? *(1 mark)*
iii) What is the density of water, in kg/m³? *(1 mark)*

b) A hot air balloon is made from a material which has a mass of 150 kg. Its volume when filled with helium is 500 m³. The density of helium is 0.18 kg/m³. Calculate the total mass of the helium-filled balloon. *(4 marks)*

5 a) i) The density of aluminium is 2.7 g/cm³. Explain what this means. *(1 mark)*
ii) Calculate the number of cm³ in 1 m³. *(1 mark)*
iii) Calculate the mass in grams of 1 m³ of aluminium. *(1 mark)*
iv) Calculate the density of aluminium in kg/m³. *(2 marks)*
b) A glass stopper weighs 40 g. It is placed in a measuring cylinder containing a liquid as shown in Figure 4.7. The cylinder gives the volume in cm³.

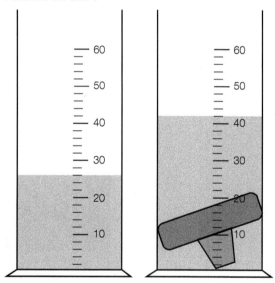

Figure 4.7

By taking readings from the diagrams, find:
i) the volume of liquid. *(1 mark)*
ii) the total volume of liquid and stopper. *(1 mark)*
iii) the volume of the stopper. *(1 mark)*
iv) the density of the stopper. *(3 marks)*

6 100 identical copper rivets are put into an empty measuring cylinder and 50 cm³ of water is added. Figure 4.8 shows the level of the water.

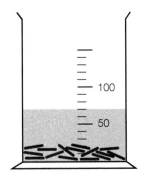

Figure 4.8

a) What is the volume of:
   i) 100 copper rivets   *(1 mark)*
   ii) 1 copper rivet?   *(1 mark)*
b) If all the copper rivets together have a mass of 180 g, calculate the density of copper.   *(1 mark)*

7 A student measured the volumes and masses of six different substances A, B, C, D, E and F. The student plotted their masses on the y-axis and their volumes on the x-axis (Figure 4.9).

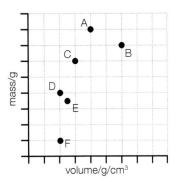

Figure 4.9

Which substances had the same density?   *(1 mark)*

8 Calculate the average density of the Earth, using the following data.

radius of Earth = 6.4×10^6 m

volume of a sphere = $\frac{4}{3} \pi r^3$

mass of the Earth = 6.0×10^{24} kg   *(5 marks)*

Energy forms

It is important to understand the difference between **energy forms** and **energy resources**. Energy forms are the different ways in which energy can appear, such as heat, light, sound, nuclear, kinetic, gravitational potential and chemical energy. Energy resources are the different ways of supplying a particular energy form. Table 5.1 summarises some of the main energy forms.

Table 5.1 Some of the main energy forms

Energy form	Definition	Examples of resources
Chemical	the energy stored within a substance, which is released on burning	coal, oil, natural gas, peat, wood, food
Gravitational potential	the energy a body contains as a result of its height above the ground	stored energy in the dam (reservoir) of a hydroelectric power station
Kinetic	the energy of a moving object	wind, waves, tides
Nuclear	the energy that is stored in the nucleus of an atom	uranium, plutonium

Other common energy forms are electrical energy, magnetic energy and strain potential energy – the energy a body has when it has been stretched or squeezed out of shape and will return to its original shape when the force is removed, such as a wind-up toy.

One of the fundamental laws of physics is the Principle of Conservation of Energy. This states that:

Energy can neither be created nor destroyed, but it can change its form.

We can show energy changes in an **energy flow diagram** (Figure 5.1).

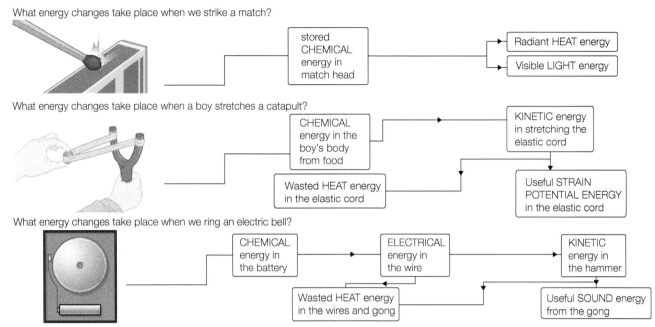

What energy changes take place when we strike a match?

| stored CHEMICAL energy in match head | → | Radiant HEAT energy |
| | | Visible LIGHT energy |

What energy changes take place when a boy stretches a catapult?

CHEMICAL energy in the boy's body from food → KINETIC energy in stretching the elastic cord

Wasted HEAT energy in the elastic cord

Useful STRAIN POTENTIAL ENERGY in the elastic cord

What energy changes take place when we ring an electric bell?

CHEMICAL energy in the battery → ELECTRICAL energy in the wire → KINETIC energy in the hammer

Wasted HEAT energy in the wires and gong

Useful SOUND energy from the gong

▲ **Figure 5.1** A variety of energy changes

Energy resources

Energy resources can be classified in several different ways. One way is to split them into renewable and non-renewable resources. Renewable resources (see Table 5.2) are those that are replaced by nature in less than a human lifetime. Non-renewable resources (see Table 5.3) are those that are used faster than they can be replaced by nature. The UK government has said that 20% of our energy needs, including 30% of the electricity we generate, must come from renewable resources by 2020.

Table 5.2 Renewable sources of energy

Renewable resource	Comment
sunlight — electric currents — solar cells — electrical components ▲ **Figure 5.2** Solar panels	Solar cells convert sunlight (solar energy) directly into electricity. Solar cells are joined together into arrays.
top lake — dam — mountain — water pumped up at night — water flows down at peak demand — step-up transformer — turbines and pump — bottom lake — National Grid ▲ **Figure 5.3** A hydroelectric power station	Because of its height from the ground, water in a dam (reservoir) contains gravitational potential energy. The water is allowed to fall from the dam through a pipe and gains kinetic energy as it falls. The fast-flowing water falls on a turbine, which then drives a generator. The output from the generator is electrical energy. Some hydroelectric power stations use pumped storage reservoirs. At times of low demand, such as in the early hours of the morning, the power station buys cheap electricity and uses it to pump water up to a high reservoir. During the day, when demand is high, they sell the electricity they produce at a higher price.

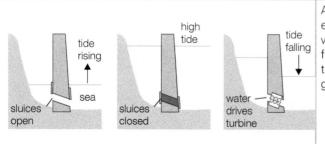

Figure 5.4 Harnessing tidal energy

A tidal barrage is created when a dam is built across a river estuary. The tide rises and falls every 12 hours, and if the water levels on each side of the dam are not equal, water will flow through a gate in the dam. The moving water drives a turbine, which is made to turn a generator. The output from the generator is electrical energy.

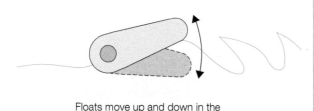

Floats move up and down in the waves to generate electricity

Figure 5.5 Generating electricity using waves

Waves are produced largely by the action of the wind on the surface of water. The wave machine floats on the surface of the water and the up and down motion of the water forces air to drive a turbine and so produces electricity.

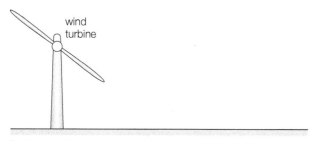

Figure 5.6 A wind turbine

As the wind blows, the large blade turns and this drives a turbine. The turbine drives a generator, which in turn produces electricity.

Large numbers of turbines are often grouped together to form a **wind farm**.

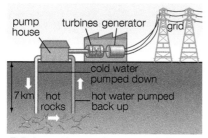

Figure 5.7 A geothermal power station

Geothermal power stations use heat from the hot rocks deep inside the Earth. Cold water is passed down a pipe to the rocks. The water is heated by these rocks and the hot water is then pumped to the surface.

Geothermal energy is often used in power stations or in district heating schemes.

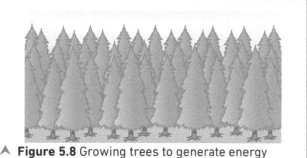

Figure 5.8 Growing trees to generate energy

Fast-growing trees, like willow, are grown on poor-quality land (or land set-aside from food production) and the timber is harvested around every three years. The wood is dried and turned into woodchips, which are then burned in power stations to produce electricity or sold for solid fuel heating.

In Brazil, biomass crops are fermented to produce alcohol. The **alcohol** is added to petrol as a way of extending the life of scarce fossil fuels. The product is called **gasohol**.
Other forms of biomass include rapeseed oil. The oil from the seeds is converted into **biodiesel** for road transport.

Table 5.3 Non-renewable sources of energy

Non-renewable resource	Comment
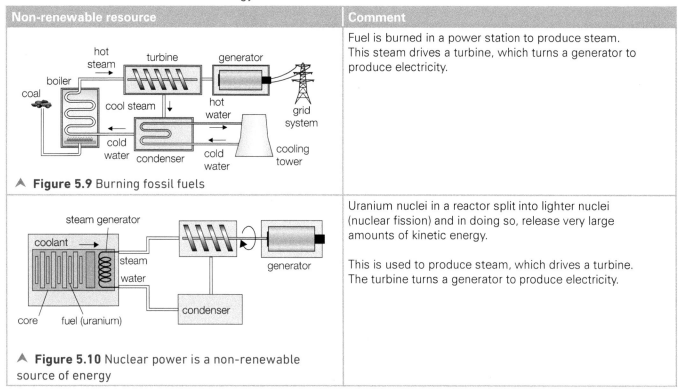 **Figure 5.9** Burning fossil fuels	Fuel is burned in a power station to produce steam. This steam drives a turbine, which turns a generator to produce electricity.
Figure 5.10 Nuclear power is a non-renewable source of energy	Uranium nuclei in a reactor split into lighter nuclei (nuclear fission) and in doing so, release very large amounts of kinetic energy. This is used to produce steam, which drives a turbine. The turbine turns a generator to produce electricity.

Advantages and disadvantages of using the different energy resources to generate electricity

Table 5.4 Advantages and disadvantages of different energy resources

Energy resource	Advantages	Disadvantages	Other comments
Fossil fuels – coal, oil, natural gas, lignite, peat	• relatively cheap to start up • moderately expensive to run • large world reserves of coal (much less for other fossil fuels).	• All fossil fuels are non-renewable. • All fossil fuels release carbon dioxide on burning, and so contribute to global warming. • Burning coal and oil also releases sulfur dioxide gas, which causes acid rain.	• Coal releases the most carbon dioxide per unit of electricity it produces, and natural gas releases the least. • Removing sulfur or sulfur dioxide is very expensive and adds greatly to the cost of electricity production.
Nuclear fuels – mainly uranium	• do not produce carbon dioxide • do not emit gases that cause acid rain.	• The waste products will remain dangerously radioactive for tens of thousands of years. • As yet, no one has found an acceptable method to store these materials cheaply, safely and securely for such a long time. • Nuclear fission fuels are non-renewable. • An accident could release dangerous radioactive material which could contaminate a very wide area, leaving it unusable for decades.	• Nuclear fuel is relatively cheap on world markets. • Nuclear power station construction costs are much higher than fossil fuel stations, because of the need to take expensive safety precautions. • Decommissioning nuclear power stations is a particularly long and expensive process, requiring specialist equipment and personnel.

Energy resource	Advantages	Disadvantages	Other comments
Wind farms	• a renewable energy resource • low running costs • reduces dependency on fossil fuels.	Wind farms are: • unreliable • unsightly • very noisy • hazardous to birds.	• Wind farms take up much more ground per unit of electricity produced than conventional power stations.
Waves	• a renewable energy resource • low running costs • reduces dependency on fossil fuels.	Wave generators at sea are: • unreliable • unsightly • hazardous to shipping.	• Many turbines are needed to produce a substantial amount of electricity.
Tides	• a renewable energy resource • low running costs • reduces dependency on fossil fuels.	Tidal barrages are built across river estuaries and can cause: • navigation problems for shipping • destruction of habitats for wading birds and the mud-living organisms on which they feed.	• Tides (unlike wind and waves) are predictable, but they vary from day to day and month to month. This makes them unsuitable for producing a constant daily amount of electrical energy.

Ireland's natural fuel resources

Ireland has almost no coal or oil resources. The island is rich in peat, but it is important not to over-exploit these resources industrially because of the damage that can be done to habitat. Ireland does, however, have an important fossil fuel resource – natural gas from the Celtic Sea. Northern Ireland has another resource – lignite. This is sometimes called brown coal because it is rocky like coal, but brown like peat. There are millions of tonnes of lignite reserves around Crumlin and under Lough Neagh.

Test yourself

1 Most of Ireland's energy needs are supplied by fossil fuels. Name three fossil fuels.
2 Make a copy of the table below and tick (✔) those items which are energy forms.

Quantity	Tick if the quantity is a form of energy
Sound	
Pressure	
Force	
Weight	
Electricity	
Heat	

3 The table below shows six energy resources. Copy the table and tick (✔) the correct boxes to show whether these resources are renewable or non-renewable.

Energy source	Renewable?	Non-renewable?
Gas		
Hydroelectricity		
Oil		
Coal		
Wind		
Tides		

4 A model aircraft has its wings covered with solar panels to drive the propellers and charge the battery. Copy and complete the following sentences to show the energy changes which take place in such an aircraft: The solar cells change .... energy into .... energy. The battery stores .... energy. As the propellers turn, they change .... energy into useful .... energy. As the model aircraft gains height, it gains .... energy. If the model aircraft crashes into the ground, it produces wasted heat and .... energy.

5 Explain what is meant by a renewable energy resource.

6 In what ways are the production of electricity in a fossil fuel power station and in a nuclear power station similar? In what ways are these power stations different?

7 Currently, nuclear waste is vitrified (turned into a type of glass), stored in strong metal drums and kept deep underground. Why is this an unsatisfactory solution for the long term?

8 Name the polluting gas that is produced by burning fossil fuels, and which contributes to global warming.

9 Norway has complained that Britain is partly responsible for the destruction of the Norwegian habitat by acid rain How might this have come about?

10 Do you think the UK government's target to have 30% of our electricity production come from renewable resources by 2020 is realistic? What can you and your family do to contribute?

11 Give three reasons for using wind farms to generate electricity.

12 An electricity company might say that electricity is a 'clean' fuel. Why is this statement misleading?

13 For each of the devices or situations shown below, use a flow diagram to show the main energy change that is taking place. The first has been done for you.

Device/situation	Input energy form	→	Useful output energy form
Microphone	sound energy	→	electrical energy
Electric smoothing iron	.... energy	→	.... energy
Loudspeaker	.... energy	→	.... energy
Coal burning in an open fire	.... energy	→	.... energy
A weight falling towards the ground	.... energy	→	.... energy
A candle flame	.... energy	→	.... energy and .... energy

Show you can ?

a) What are the arguments for and against installing a nuclear power station in Ireland?

b) Imagine you are a government scientist. Write about 100 words giving the advantages of building a nuclear power station rather than one which burns fossil fuels.

c) Why do you think Northern Ireland has not yet mined the lignite resources around Crumlin?

The Sun

Almost all energy resources ultimately rely on the energy of the Sun. In the case of fossil fuels, we know that these resources come from the dead remains of plants and animals laid down many millions of years ago. The plants obtained their energy from the Sun by **photosynthesis**. Herbivores ate the plants, and carnivores ate the herbivores. Under the Earth's surface, these remains slowly fossilised into coal, peat, gas and oil.

Other processes also rely on the Sun's energy. Hydroelectric energy depends on the water cycle, and this process begins when ocean water evaporates as a result of absorbing radiant energy from the Sun. Wind and waves rely on the Earth's weather, which is largely controlled by the Sun. Only geothermal and nuclear energy do not depend directly on the energy emitted by the Sun.

1 An electric kettle is rated 2500 W. It produces 2500 J of heat energy every second. The kettle takes 160 seconds to boil some water, and during this time, 360 000 J of heat energy pass into the water. Find the kettle's efficiency.

Answer

useful energy output (passed into water) = 360 000 J

total energy input = 2500 × 160 = 400 000 J

$$\text{efficiency} = \frac{\text{useful energy output}}{\text{energy input}}$$

$$= \frac{360\,000}{400\,000}$$

$$= 0.9$$

Therefore:

- 90% of the electrical energy is used to boil the water
- 10% of the energy supplied is wasted

Most will be passed through the kettle as wasted heat to the surrounding air. A small amount of heat will be lost as some water evaporates.

2 A motor rated 40 W lifts a load of 80 N to a height of 90 cm in 4 s. Find its efficiency.

Answer

useful energy output

= work done by motor

= force × distance

= 80 × 0.9

= 72 J

$$\text{power} = \frac{\text{energy supplied}}{\text{time taken}}$$

$$40 = \frac{\text{energy supplied}}{4}$$

40 × 4 = energy supplied

160 J = energy supplied

$$\text{efficiency} = \frac{\text{useful energy output}}{\text{energy input}}$$

$$= \frac{72}{160}$$

$$= 0.45$$

Efficiency

Efficiency is a way of describing how good a device is at transferring energy from one form to another in an intended way.

If a light bulb is rated 100 W, this means that it normally uses 100 J of electrical energy every second. But it might only produce 5 J of light energy every second (see pages 53–54). The other 95 J are wasted as heat. This means that only 5% of the energy is transferred from electrical energy into light energy. This light bulb therefore has an efficiency of 0.05, or 5%. If the same light bulb were used as a heater, its efficiency would be 95%, because the intended output energy form would be heat, not light.

Efficiency is defined by the formula:

$$\text{efficiency} = \frac{\text{useful energy output}}{\text{energy input}}$$

As efficiency is a ratio, it has no units. By the Principle of Conservation of Energy, energy cannot be created, so the useful energy output can never be greater than the energy input. However, energy is wasted in every physical process, so the efficiency of a machine is always less than 1.

14 The electrical energy used by a boiler is 1000 kJ. The useful output energy is 750 kJ.
   a) Calculate the efficiency of the boiler.
   b) Suggest what might have become of the energy wasted by the boiler.

15 Explain why the efficiency of a device can never be greater than 1.00 or 100%.

16 A car engine has an efficiency of 0.28. How much input chemical energy must be supplied if the total output of useful energy is 140 000 kJ?

17 Figure 5.11 shows energy transfers in a mobile phone.

▲ **Figure 5.11** Energy transfers in a mobile phone

   a) Use the figures on the diagram to calculate the phone's efficiency.
   b) What principle of physics did you use to calculate the useful sound energy produced?

18 Figure 5.12 shows a rotary engine which has an efficiency of 30%.
   a) Calculate the amount of useful energy it produces when the input chemical energy is 2000 J.
   b) 90% of the wasted energy is heat. What percentage of the input energy is lost as heat?

▲ **Figure 5.12** Energy efficiency in a rotary engine

Work

Work is only done when a force causes movement. Although pushing against a wall might make a person tired, no work is done on the wall because it produces no movement. Similarly, holding a book at arm's length is doing no work on the book. Lifting a book from the floor and placing it on a table is doing work because we are applying a force and producing movement.

We can calculate work using the following formula:

work done = force × distance moved in direction of force or:

$$W = F \times d$$

The units in this formula are matched. Force must always be measured in **newtons**, but if the distance were in cm, the work would be in N cm. If the distance were in metres, the work would be in N m. The N m occurs so often that physicists have renamed it the **joule** (J).

The joule is defined as the energy transferred when the point of application of a force of 1 newton moves through one metre.

Doing work means 'spending' energy. The more work a person does, the more energy they need. The energy used is equal to the amount of work done.

Example

1 How much work is done when a packing case is dragged 4 m across the floor against a frictional force of 45 N? How much energy is needed?

Answer

Since the case moves at a steady speed, the forward force must be the same size as the friction force. So the forward force is 45 N.

$$\text{So } W = F \times d$$
$$= 45 \times 4$$
$$= 180 \text{ J}$$

Energy needed = work done
= 180 J

2 A crane does 1200 J of useful work when it lifts a load vertically by 60 cm. Find the weight of the load.

Answer

Since the load is being lifted, the minimum upward force is the weight of the load.

$$W = F \times d$$
$$1200 = F \times 0.6 \text{ (convert 60 cm to 0.6 m)}$$
$$F = 1200 \div 0.6$$
$$= 2000 \text{ N}$$

Hence, weight of load = 2000 N

3 How much work is done by an electric motor pulling a 130 N load 6.5 m up the slope shown in Figure 5.13 if the constant tension in the string is 60 N?

Answer

Since the tension and distance moved are both parallel to the slope, they are both used to find the work done. The weight of the load is not used in this question.

work = force × distance moved in direction of the force
= 60 × 6.5
= 390 J

tension = 60 N, distance = 6.5 m

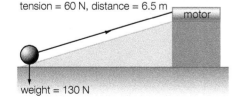

weight = 130 N

▲ **Figure 5.13**

19 Competitors in a strength competition must throw a cement block of mass 100 kg over a wall that is 5.5 m high. How much work is done if the block just clears the top of the wall?

20 A man pushes a lawn mower with a force of 60 N. How much work does he do when he pushes the lawn mower 20 m?

21 Stephen weighs 550 N. How much work does he do in climbing up to a diving board which is 3.0 m high?

Example

1 An electric motor is used to raise a load of 105 N. The load rises vertically 2 m in a time of 6 s. Find the work done and the power of the motor.

Answer

To lift the load, the motor must produce an upward force of at least 105 N. work done = force × distance

$$= 105 \times 2$$
$$= 210 \text{ J}$$

$$\text{power} = \frac{\text{work done}}{\text{time taken}}$$
$$= \frac{210}{6}$$
$$= 35 \text{ W}$$

2 A crane has a power of 2000 W. How much work could it do in an hour?

Answer

In power calculations, the unit of time is the second.

So first convert 1 hour to seconds:

1 hour = 60 minutes
$$= 60 \times 60 \text{ seconds}$$
$$= 3600 \text{ seconds}$$

$$2000 = \frac{\text{work done}}{3600}$$
work done = 2000 × 3600
$$= 7\,200\,000 \text{ J or 7.2 MJ}$$

Work and energy

Energy is the ability to do work. If a machine has 500 J of stored energy, this means it can do 500 J of work. Similarly, work is sometimes thought of as the amount of energy transferred. Note that both work and energy are measured in joules.

Tip

In all calculations of this type, first write down the appropriate formula, then substitute the values you know. Rearrange the formula if necessary, and carry out the calculations with a calculator. Give your final answer with its unit. Remember to always show your working.

Example

A battery stores 15 kJ of energy. If the battery is used to drive an electric motor, how high could it raise a 750 kg load if it was lifted vertically?

Answer

The battery stores 15 kJ or 15 000 J, so it can do a maximum of 15 000 J of work.

Since a mass of 1 kg weighs 10 N, a mass of 750 kg has a weight of 7500 N. The motor must therefore produce an upward force of at least 7500 N.

$$W = F \times d$$
$$15\,000 = 7500 \times d$$
$$d = 15\,000 \div 7500$$
$$= 2 \text{ m}$$

Note that 2 m is the highest that this motor could raise the load. It is likely that it would not raise the load quite this high because some of the energy in the battery is used to produce heat and sound. In our calculation, we have assumed that all the energy in the battery is used to do work against gravity.

Power

Power is the amount of energy transferred in one second, or the amount of work done in one second.

This means that the power of a machine is the work it can do in a second.

The formula for calculating power is:

$$\text{power} = \frac{\text{work done}}{\text{time taken}} \text{ or } P = \frac{W}{t}$$

Work is measured in joules and time is measured in seconds, so power must be measured in joules per second or J/s. The J/s is also known as the **watt (W)**, named after James Watt, the Scottish engineer.

$$1 \text{ W} = 1 \text{ J/s}$$

More generally, power may also be defined as the rate of change of energy transferred.

$$\text{power} = \frac{\text{energy transferred}}{\text{time taken}}$$

22 A person weighing 550 N runs up the stairs in 3 seconds. The stairs are made of 15 steps each of 14 cm height. Find the person's average power.

23 A nail gun fires a nail with a kinetic energy of 1.8 J into a piece of wood. The average resistive force on the nail is 45 N, and it stops 0.3 s after entering the wood.
Calculate:
a) the distance the nail penetrates into the wood
b) the average power of the resistive forces in stopping the nail.

Prescribed practical

Investigating personal power

▲ **Figure 5.14** How to measure personal power

Aims
- to measure the weight of a student
- to measure the height of stairs
- to calculate personal power

Apparatus
- bathroom scales
- metre ruler
- stopwatch

Method
1 Prepare a table of results similar to that shown in Table 5.5.
2 Measure mass of a student using bathroom scales.
3 Convert mass to weight using the formula $W = mg$.
4 Find the height of one riser (step) in metres, using a metre rule.
5 Count the number of risers and multiply by the height of one riser. This is the total vertical height of the stairs.
6 Ask another student to measure the time to run up the stairs, using a stopwatch.
7 Repeat and work out the average time for running up the stairs.
8 Use the formulae for work and power to calculate the personal power of the student.

Results
Table 5.5

Mass of student/kg	45
Weight of student/N	450
Height of risers/cm	14.0, 13.8, 13.8, 14.0, 13.9
Average riser height/cm	13.9
Number of risers	30
Vertical height of staircase	13.9 × 30 = 417 cm = 4.17 m
Time to run upstairs/s	5.0

Calculations

$$\text{work done} = \text{force} \times \text{distance}$$
$$= 450 \times 4.17$$
$$= 1876.5 \, \text{J}$$
$$\text{power} = \frac{\text{work done}}{\text{time taken}}$$
$$= \frac{1876.5}{5.0}$$
$$= 375.3 \, \text{W}$$

Conclusion
The average power was found to be............

Practical activity

Power of an electric motor

Figure 5.15 shows the apparatus that can be used to determine the power of an electric motor.

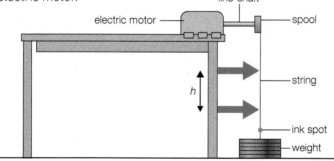

▲ **Figure 5.15** Determining the power of an electric motor

1 Attach a weight (or mass) to end of string that is attached to motor shaft so that when the motor is switched on, the weight is raised vertically.
2 Measure the distance, h, between two pointers or markers using a metre ruler.
3 Switch on the motor and use a stopwatch to measure the time taken for the spot on the string to move between the two pointers.
4 Repeat and find an average time, to improve reliability and accuracy.

$$\text{work done} = \text{force} \times \text{distance moved}$$
$$\text{power} = \frac{\text{work done}}{\text{time taken}}$$

Extension

The experiment could be repeated using different weights. You should find that the power of the motor is not constant when the weight is changed.

Table 5.6 Typical results for a small motor

Weight/N	Height h/m	Work done/J	Time taken 1/s	Time taken 2/s	Time taken 3/s	Average time/s	Average power/W
9.0	1.8	16.2	5.4	5.5	5.3	5.4	3.0
10.0	1.8	18.0	5.6	5.6	5.6	5.6	3.2
11.0	1.8	19.8	6.0	6.1	5.9	6.0	3.3
12.0	1.8	21.6	6.5	6.5	6.5	6.5	3.3
13.0	1.8	23.4	7.3	7.5	7.1	7.3	3.2

Validity of data

Data is only **valid** if the measurements taken are affected by a **single independent variable** only. Data is not valid if, for example, a **fair test** is not carried out, or there is observer bias.

In this case, there is only one independent variable (the weight being lifted by the motor) and the test is fair. The height and the input voltage are constant; they are controlled variables to ensure that the test is fair.

There should be no cause here to suspect observer bias, but just to be sure, professional scientists like to repeat each other's experiments to check that they get similar results.

Reliability of data

To ensure the results are **reliable**, the experiment is repeated several times and the average power is determined for different weights.

Kinetic energy

The kinetic energy (KE) of an object is the energy it has because it is moving. It can be shown that an object's kinetic energy is given by the formula:

kinetic energy = ½ × mass × velocity² or KE = ½ mv^2

where m is the mass in kg and v is the speed of the object in m/s.

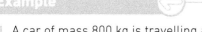

Example

1 A car of mass 800 kg is travelling at 15 m/s. Find its kinetic energy.

Answer

KE = ½ mv^2

= ½ × 800 × 15²

= 90 000 J

2 A bullet has a mass of 20 g and is travelling at 300 m/s. Find its kinetic energy.

Answer

20 g = 0.02 kg

KE = ½ mv^2

= ½ × 0.02 × 300²

= 900 J

3 Find the speed of a boat if its mass is 1200 kg and it has a kinetic energy of 9600 J.

Answer

KE = ½ mv^2

= ½ × 1200 × v^2

$v^2 = \dfrac{9600}{600}$

$v = 4$ m/s

4 The input power of a small hydroelectric power station is 1 MW.
If 18 000 000 kg of water flows past the turbines every hour, find the average speed of the water.

Answer

1 hour = 60 × 60 seconds

= 3600 seconds

Since a 1 MW power station produces 1 000 000 J of electrical energy per second, the minimum KE of the water passing every second is 1 000 000 J.

KE = ½ mv^2

1 000 000 = ½ × 5000 × v^2

$v^2 = \dfrac{1\,000\,000}{2500} = 400$

$v = 20$ m/s

Gravitational potential energy

Test yourself

24 A communications satellite of mass 120 kg orbits the earth at a speed of 3000 m/s. Calculate its kinetic energy.

25 The viewing platform at the Eiffel tower in Paris is about 280 m from the ground. Find the gravitational potential energy of a rubber of mass 50 g on the viewing platform. Compare this to the kinetic energy of a 10 g bullet travelling at 150 m/s. Comment on your answer.

26 An oil tanker has a mass of 100 000 tonnes. Its kinetic energy is 200 MJ. Calculate its speed.
(1 tonne = 1000 kg, 1 MJ = 1 000 000 J)

27 A car of mass 800 kg is travelling at a steady speed. The kinetic energy of the car is 160 000 J. Show carefully that the speed of the car is 72 km/h.

When any object with mass is lifted, work is done on it against the force of gravity. The greater the mass of the object and the higher it is lifted, the more work has to be done. The work that is done is only possible because some energy has been transferred. This energy is stored in the object as gravitational potential energy (GPE).

When the object is released, it falls back to Earth and the stored energy can be recovered. If the object crashes into the ground, a bang (sound energy) is heard and heat is produced.

Gravitational potential energy is the work done raising a load mass m, against the force of gravity (g) through a height (h), so:

$$GPE = mgh$$

where m is the mass in kg, g is the gravitational field strength in N/kg and h is the vertical height in m.

It is important to remember that 1 kg has a weight of 10 N on the surface of the Earth. This is just another way of saying that the gravitational field strength, g, on Earth is about 10 N/kg.
The value of g is different at different parts of the Universe.
For example, g on the Moon is only about a sixth of its value on Earth, approximately 1.6 N/kg.

> **Tip**
>
> If a process had just GPE at the start and just KE at the end, you know that the GPE must equal the KE.

Example

1 Find the gravitational potential energy of a mass of 500 g when raised to a height of 240 cm.
Take g = 10 N/kg
500 g = 0.5 kg
240 cm = 2.4 m

Answer
GPE = mgh
= 0.5 × 10 × 2.4
= 12 J

2 How much heat and sound energy is produced when a mass of 1.2 kg falls to the ground from a height of 5 m? Take g = 10 N/kg.

Answer
Heat and sound energy produced
= original GPE
= mgh
= 1.2 × 10 × 5
= 60 J

3 How much gravitational potential energy is stored in the reservoir of a hydroelectric power station if it holds 5 000 000 kg water at an average height of 80 m above the turbines?

Answer
GPE = mgh
= 5 000 000 × 10 × 80
= 4 000 000 000 J

4 A marble of mass 50 g falls to the Earth. At the moment of impact, its kinetic energy is 1 J. From what height did it fall?

Answer
50 g = 0.05 kg
KE at impact = GPE at start
1 = mgh
= 0.05 × 10 × h
= 0.5 × h
h = 2 m

5 A book of mass 500 g has a gravitational potential energy of 3.2 J when at a height of 4 m above the surface of the Moon. Find the gravitational field strength on the Moon.

Answer
GPE = mgh
3.2 = 0.5 × g × 4
= 2g
g = 3.2 ÷ 2
= 1.6 N/kg

28 A ball of mass 2 kg at rest then falls from a height of 5 m above the ground. Copy the table below and complete it to show the gravitational potential energy, the kinetic energy, speed and the total energy of the falling ball at different heights above the ground.

Height above ground/m	Gravitational potential energy/J	Kinetic energy/J	Total energy/J	Speed/m/s
5.0		0	100	0
4.0				4.47
	64			
1.8		64		
0.0	0			

29 The power of the motor in an electric car is 3600 W. How much electrical energy is converted into other energy forms in 5 minutes?

30 A crane can produce a maximum output power of 3000 W. It raises a load of mass 1500 kg through a vertical height of 12 m at a steady speed.
  a) i) What is the weight of the load?
     ii) How much useful work does the crane do lifting the load 12 m?
  b) How long does it take the crane to raise the load 12 m?
  c) At what speed will the load rise through the air?

31 A barrel of weight 1000 N is pushed up a ramp. The barrel rises vertically 40 cm when it is pushed 1 m along the ramp.
  a) Calculate how much useful work is done when the barrel is pushed 1 m along the ramp.
  b) To push the barrel 1 m along the ramp requires 1200 J of energy. Calculate the efficiency of the ramp.

32 On planet X, an object of mass 2 kg is raised 10 m above the surface. At that height, the object has a gravitational potential energy of 176 J. Details of three planets are given below. Which one of these three planets is most likely to be planet X?

Planet	Mercury	Venus	Jupiter
Gravitational field strength, g/N/kg	3.7	8.8	26.4

33 A bouncing ball of mass 200 g leaves the ground with a kinetic energy of 10 J.
  a) If the ball rises vertically, calculate the maximum height it is likely to reach.
  b) In practice, the ball rarely reaches the maximum height. Explain why this is so.

1 A satellite orbits the Earth. Name the two main types of energy possessed by the satellite in its orbit. *(2 marks)*

2 Electricity can be generated by wind turbines.
   a) Copy and complete the sentences below to show the useful energy change which takes place in a wind turbine.
   .... energy of the wind is transferred to .... energy. *(2 marks)*
   b) The wind is a renewable energy source. What does this mean? *(1 mark)*
   c) Give two other examples of renewable energy resources. *(2 marks)*

3 In Scotland, hydroelectric power makes a significant contribution as a source of electricity (Figure 5.16).

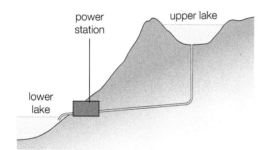

power station   upper lake
lower lake

Figure 5.16

Copy and complete the flow diagram in Figure 5.17 to show the energy changes taking place in a hydroelectric power station.

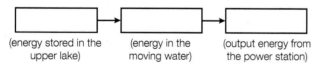

(energy stored in the upper lake) → (energy in the moving water) → (output energy from the power station)

Figure 5.17 *(3 marks)*

4 A tidal barrage in France generates electricity. One environmental effect of using the tides to generate electricity is that it reduces global warming by decreasing the consumption of fossil fuels.
   a) Explain fully how this reduces global warming. *(2 marks)*
   b) Apart from the above environmental issue, state one advantage and one disadvantage of generating electricity from the tides. *(2 marks)*

5 a) The most common energy resources used in Europe today are oil, natural gas, coal, nuclear energy, hydroelectric and wind energy.
   i) Choose one non-renewable energy resource from the list above and say why it is non-renewable. *(2 marks)*
   ii) Choose one renewable energy resource from the list above and say why it is renewable. *(2 marks)*
   iii) Give one advantage that non-renewable energy resources have over renewable energy resources. *(1 mark)*
   b) It has been estimated that 1×10^8 kg (100 000 000 kg) of water flows over Niagara Falls every second. The falls are 50 metres high.
   i) Calculate the gravitational potential energy lost every second by the water flowing over the falls ($g = 10$ m/s^2). *(3 marks)*
   A feasibility study has shown that only 0.008 (0.8%) of the available potential energy could be converted into electrical energy by a hydroelectric power station built on the falls.
   ii) Calculate the maximum power output of such a hydroelectric power station. *(3 marks)*
   iii) Explain why all hydroelectric power stations are dependent on the energy of the Sun. *(2 marks)*
   c) Figure 5.18 shows a vehicle with a winch attached. The winch is connected to a tree by rope. As the winch winds in the rope, the vehicle moves forward towards the tree. The winch uses 500 W of input electrical power. It has an efficiency of 0.6.

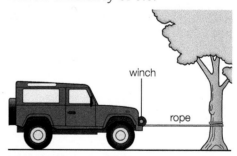

winch   rope

Figure 5.18
   i) Calculate the useful output power of the winch. *(3 marks)*
   ii) Write down the useful work done by the winch in 1 second. *(1 mark)*
   iii) The pulling force in the rope is 1200 N. Calculate the constant speed at which the vehicle moves forward. *(3 marks)*

6 a) How much work is done by a tractor when it lifts a load of 8000 N to a height of 1.8 m?

(3 marks)

b) The output power of the tractor is 5.2 kW. How long does it takes to do 26 000 J of work?

(3 marks)

c) The efficiency of the tractor is 0.26 (26%). If the output power of the tractor is 5.2 kW, calculate the input power.

(3 marks)

7 Saltburn is a seaside resort in Yorkshire. There is a considerable drop from the cliff top to the beach. In 1884, an inclined tramway was built to carry passengers from the beach to the cliff top. Two identical tramcars were used, each with a water tank underneath. The tramcars were connected by a steel cable which passed around a large pulley at the top (Figure 5.19). The tramcar that happens to be at the top has water added until there is enough to raise the tramcar at the bottom of the tramway.

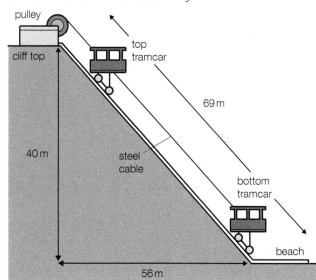

Figure 5.19

a) On one journey, the weight of the lower tramcar and its passengers was 24 000 N. Ignoring friction, calculate the work done, in kJ, to bring the tramcar from the beach to the cliff top.

(3 marks)

b) The time for this journey was 20 seconds. Calculate the power needed to raise the tramcar.

(3 marks)

c) On this journey, the energy provided by the upper car as it descended was 1200 kJ. Calculate the efficiency of the tramway on this journey.

(3 marks)

d) During the journey, certain energy changes take place. Copy and complete the following table by stating whether the energy listed in the first column increases, decreases or remains unchanged as the top tramcar descends at a constant speed.

Energy	Increases/decreases/unchanged
Potential energy of the top tramcar	
Kinetic energy of the top tramcar	
Kinetic energy of the bottom tramcar	
Potential energy of the bottom tramcar	
Heat energy	

(5 marks)

8 a) A basketball player throws a ball up into the air. Copy the table and place a tick (✓) in the appropriate column of your table to show what happens to each quantity as the ball rises. Ignore the effects of friction.

Quantity	Increases	Decreases	Remains constant
Speed of the ball			
Potential energy of the ball			
Total energy of the ball			
Potential energy of the bottom tramcar			
Kinetic energy of the ball			

(4 marks)

b) A heavy ball, of mass 10 kg, is dropped from a height of 5 metres.

i) What is the potential energy lost by the ball during this fall?

(3 marks)

ii) Calculate the velocity of the ball at the bottom of the fall.

(4 marks)

9 A ball of mass 3 kg is dropped from the top of a tall building. The ball loses 60 J of energy due to air resistance on its way down. When it strikes the ground, it has a kinetic energy of 600 J.

a) Calculate the gravitational potential energy of the ball when it is released from the top of the building.

(1 mark)

b) Calculate the height of the building.

(3 marks)

c) Calculate the speed of the ball before it strikes the ground.

(3 marks)

10 A boulder, of mass 440 kg, rolls down a slope and into the sea. At the edge of the cliff, the boulder has a kinetic energy of 3520 J and a potential energy of 52 800 J.
   a) Calculate the height of the cliff. *(3 marks)*
   b) Calculate the kinetic energy of the boulder as it strikes the water. Assume no energy losses. *(1 mark)*
   c) Calculate the velocity of the boulder when it hits the water. *(4 marks)*

11 A crane uses a wrecking ball to demolish an old building. The diagram shows the motion of the ball. The crane moves the ball from its rest position X up to position Y, where it comes momentarily to rest before falling to collide with the wall of the building.

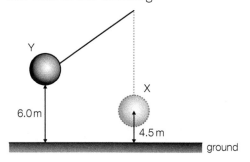

Figure 5.20
   a) Calculate the work done on the ball to move it from its rest position X to position Y. The ball has a mass of 800 kg. *(3 marks)*
   b) What is the loss in potential energy of the ball as it swings from position Y to position X? *(1 mark)*
   c) For the ball to be effective, it must have a minimum kinetic energy of 4900 J. Calculate the velocity of impact for this energy. *(4 marks)*

12 Look at Figure 5.21. A toy car of mass 1.2 kg is released from rest at point A, before it 'loops the loop'.

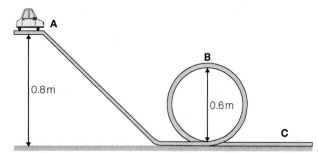

Figure 5.21
   a) Calculate the difference in potential energy of the toy car between points A and B. *(4 marks)*

b) Calculate the velocity of the toy car at point C, if its kinetic energy at C is 4.2 J. Assume no energy losses due to friction. *(2 marks)*

13 a) Figure 5.22 shows a solar panel. This is made up of a number of photocells. The photocells produce electricity directly from sunlight. Solar panels are placed on the roof of a house. On a cloudless summer day, the solar energy shining on the panel every second is 6000 J. Of this amount, 4800 J are reflected, and the rest is converted to electricity.

Figure 5.22
   i) Calculate the output of electrical energy every second from the solar panel. *(1 mark)*
   ii) Calculate the efficiency of the solar panel. *(2 marks)*
   iii) On a certain summer day, the panel generated electricity for 10 hours. Calculate the number of kilojoules generated on this day by the solar panel. *(2 marks)*
   iv) State one advantage and one disadvantage of using solar panels. *(2 marks)*
   v) A family of four would use on average 54 000 kJ of electrical energy per day. State two things they could do to make up the difference between what the solar panel produces and what they need. *(2 marks)*
b) John uses a weights machine in a gym.
   i) When using the machine, John wants to do 300 J of work in each lift. He can vary the weight from 100 N to 500 N in steps of 50 N. He can also vary the distance he lifts the weights from 1.0 m to 2.0 m in steps of 0.5 m.
       State three weights and the corresponding distances that John can use to achieve this.
       1. weight =   distance =
       2. weight =   distance =
       3. weight =   distance =
       (3 marks)
   ii) John repeats the exercise. He does 10 complete lifts in a time of 30 seconds. Calculate the power John produces during this time. *(3 marks)*

6 Heat transfer

Specification points

This chapter covers sections 1.4.22 to 1.4.25 of the GCSE Physics specification. Note: This chapter is **not** in the Double Award specification. It covers the basic methods of heat transfer and ways of reducing heat loss in the home.

There are three main methods of heat transfer.

1 Conduction – occurs mainly in solids. Most liquids are very poor conductors of heat and almost no heat conduction takes place in gases. For this reason, trapped air in fibreglass wool is an excellent insulator.

2 Convection – transfers heat only in liquids and gases.

3 Radiation – the only method of heat transfer in a vacuum.

Conduction

Conduction in metals

Most metals are good conductors of heat. Figure 6.1 shows an experiment to test which of four materials is the best heat conductor. The four rods rest on a tripod, and a small pin is attached to each one using candle wax as 'glue'. The other ends of the rods are heated equally with a Bunsen flame. To make it a fair test, all the rods must be the same length and have the same cross-sectional area.

Heat conducts along all of the rods, but the pins fall off at different times. The pin attached to the copper rod falls first, shortly followed by the pin attached to the aluminium rod, then the pin attached to the steel rod. Only after many minutes does the pin attached to the glass rod fall.

The experiment shows that copper and aluminium are very good conductors, steel is a good conductor, but glass is a very poor heat conductor. Poor conductors are called insulators.

Conduction in insulators such as glass

In solids, the atoms are held together by chemical bonds. Although they cannot move freely around within the solid, they are able to vibrate. The part of the glass that is in the flame absorbs heat energy (Figure 6.2). This makes the atoms in the end of the rod vibrate faster and with greater amplitude than their neighbours. These vibrations pass from atom to atom through the solid structure, transferring heat in the form of kinetic energy as they do so. Only after a considerable time does the energy of the flame reach the other end of the glass rod.

copper rod
steel rod
glass rod
aluminium rod
drawing pin

▲ **Figure 6.1** This apparatus can be used to demonstrate heat conduction in different materials

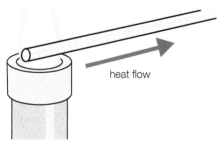

heat flow

▲ **Figure 6.2** In a glass rod, heat is conducted slowly as the vibrations pass from one atom to the next

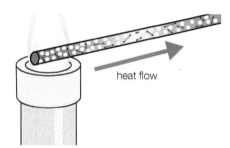

▲ **Figure 6.3** In a metal rod, heat is conducted rapidly through the movement of free electrons

▲ **Figure 6.4** Inside the reactor in this nuclear submarine, liquid sodium is used to conduct heat away from the core of the reactor

Free electron conduction

Why is copper such a good conductor of heat? Unlike glass, copper has free electrons in its metallic structure. These are electrons that have escaped from atoms and can move freely throughout the solid. The free electrons absorb heat from the Bunsen flame (Figure 6.3). This heat allows them to move much faster than before. As they move through the metal, free electrons collide with copper atoms. In these collisions, the electrons give some of their kinetic energy to the atoms and cause them to vibrate with greater amplitude than before. Free electron conduction is much faster than conduction caused by passing vibrations from atom to atom, so materials with free electrons (including all metals) are the best conductors of heat energy.

This explains why liquids are such poor conductors. Most liquids have no free electrons, so they rely on passing vibrations from atom to atom. But because the atoms in a liquid are constantly moving, conduction cannot take place in an orderly way. Liquid metals are exceptions to this rule. At room temperature, mercury is a liquid, but it is also a metal, so it conducts heat using its free electrons. This is one of the reasons why it was used for many years in thermometers.

Liquid sodium (another liquid metal) is used to conduct the heat away from the reactor core in nuclear submarines (Figure 6.4). The heat conducted from the reactor is then used to produce electricity to power the submarine below the waves.

Why does the metal blade of a knife always feel colder than the plastic handle when taken from a drawer (Figure 6.5a)? The first thing to understand is that the handle and the blade are at the same temperature, and your hand is warmer than both of them. Because the metal blade is a better conductor than the plastic handle, the blade conducts heat away from your hand faster than the plastic, making it feel colder. Touching very cold metals can cause heat to be conducted from your hand so rapidly that it can give you a serious 'burn'.

The reverse happens if a knife is removed from hot water (Figure 6.5b). As the blade is hotter than your skin, heat is conducted from the blade into your hand faster than heat from the handle. This makes the blade feel hotter than the handle.

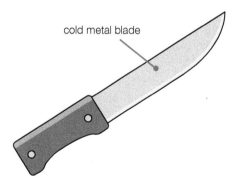

▲ **Figure 6.5a** When the knife is at room temperature, the metal blade will feel colder than the handle

▲ **Figure 6.5b** When the knife is hotter than the hand, the metal blade will feel hotter than the handle

Container ships are used to carry fruit and vegetables all over the world. The hold of the ship has two metal walls with an insulator between them.

a) What is the purpose of the insulating material?

b) Give the name of a suitable insulator for this purpose.

c) What makes this insulator effective?

Air – nature's insulator

What happens when we hold a match a few centimetres away from a flame (Figure 6.6)? The heat reaching the match head is not enough to light the match. This is because air is a very poor conductor of heat.

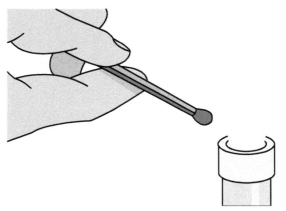

▲ **Figure 6.6** A match placed a short distance away from a flame will not light

Convection

Convection in air

Convection occurs when the fastest-moving particles in a hot region of a gas or liquid move to a cool region, taking their heat energy with them. It occurs only in liquids and gases, because the atoms in solids are not free to move from place to place. Convection is explained by changes in the density of the material.

Convection in the air can be demonstrated by the glass chimney experiment shown in Figure 6.7. First, a straw is lit and the flame is blown out. Note that the smoke rises when the straw is held in the air. Then the smoky straw is held over each chimney in turn.

When the smoky straw is over the candle flame, the smoke rises.

When the straw is held over the other chimney, the smoke falls! If the straw is held in position for long enough, the smoke will eventually be seen in the horizontal section, and then rising above the candle flame.

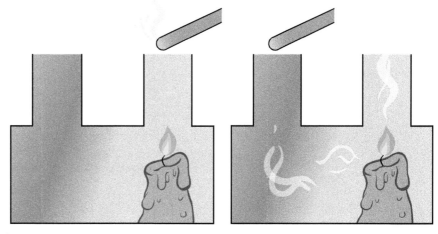

▲ **Figure 6.7** Demonstrating convection currents in air

Why does the smoke fall down the chimney?

▶ The air around the candle flame becomes very hot.

▶ The air molecules near the flame are moving faster than those in normal air.

▶ The hot air molecules are further apart than those in normal air because the air has expanded.

▶ The density of the hot air is less than that of normal air, so the hot air rises up the chimney.

▶ Cooler air moves along the horizontal tunnel to replace the air that has gone up the chimney.

▶ The moving smoke follows the motion of the cooler air.

Convection in a liquid

Tip

When explaining convection, you must not write 'heat rises'.

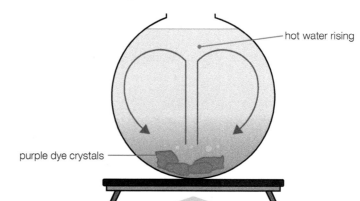

▲ **Figure 6.8** Demonstrating convection currents in a liquid

Figure 6.8 shows convection in water.

The movement of the purple dye in the water shows the convection current.

As the water at the bottom of the flask warms up, the molecules gain kinetic energy. This extra energy causes the following to happen:

▶ The molecules vibrate with greater amplitude.

▶ This causes the warm water to expand.

▶ The density of the warm water is less than that of the cold water.

▶ The warm water rises.

▶ The cooler water flows downwards to replace the upward-moving warmer water.

▶ The cool water at the top falls as it is replaced by warm water.

Reducing heat loss from your home

Heat is lost through the roof, walls, windows and floor of a home, as shown in Figure 6.9. Different materials and devices have been designed to reduce this heat loss.

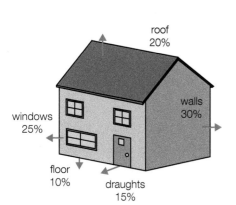

▲ **Figure 6.9** Percentages of total heat loss from different parts of the house

Use the information in Table 6.1 to help you to answer these questions.

a) What is meant by saying that the time taken to earn back the cost of double glazing is 40 years? Suggest a reason why people have their windows double glazed.

b) Which of the methods of reducing heat loss in the home would you recommend to a young family on a limited budget? Give reasons for your answers.

Table 6.1 Ways and costs of reducing heat loss in your home

Device	Time taken to earn back cost using money saved on heating bills	How losses are reduced
Cavity wall insulation	3 years	The cavity between the outside walls is filled with fibreglass, mineral wool or foam. These materials trap air in tiny pockets. Trapped air reduces heat loss through walls by convection and conduction.
Loft (attic) insulation	1.5 years	Fibreglass or mineral wool fibres trap air. Trapped air reduces heat loss through the roof by convection and conduction.
Double glazing	40 years	Two panes of glass are placed in a frame with a small gap between them. The trapped air reduces heat loss by convection and conduction.
Thick curtains and carpets	Variable (depends on quality)	Reduces draughts. Trapped air reduces heat loss through windows and floors by convection and conduction.

Radiation

Radiation is the method of heat transfer that takes place without the need for any particles. It is the way by which the Earth receives heat energy from the Sun through the vacuum of space. The heat energy is transferred as **infrared radiation**, which is part of the electromagnetic spectrum.

All objects radiate energy (emit radiant heat). The hotter an object is, the more radiation it emits. All objects also absorb radiant heat. If an object is hotter than its surroundings, it emits more radiant heat than it absorbs, so its temperature falls. If an object is cooler than its surroundings, it absorbs more radiant heat than it emits, so its temperature rises.

Giving out radiation (radiation emission)

Figure 6.10 shows an experiment in which a thick piece of copper is covered with gloss (shiny) white paint on one side and matt (non-shiny) black paint on the other. The copper has been heated with a Bunsen burner until it is very hot.

If you were to hold your hand about 30 cm from the gloss white side, and then hold your hand about the same distance from the black side, you would notice that your hand would feel much hotter facing the matt black surface. This is because the matt black surface is the better emitter of radiant heat.

Rules to remember:
▶ Black surfaces are the best emitters of radiation.
▶ White surfaces are the worst emitters of radiation.
▶ Matt surfaces are better emitters than gloss surfaces.

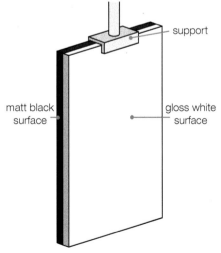

▲ **Figure 6.10** Demonstrating radiation from two different types of surface

matt black surface — support — gloss white surface

Taking in radiation (radiation absorption)

Figure 6.11 shows two sheets of thin aluminium, one painted gloss white and the other matt black. A cork is fixed to the back of each vertical plate with candle wax as 'glue'. The plates are placed equal distances away from a Bunsen burner. When the burner is lit, each plate receives the same quantity of radiant heat. The wax on the matt black plate will melt first, and the cork will fall off. The white plate warms much more slowly, so the cork takes much longer to fall off. This is because the black surface is the better absorber and the white surface is the better reflector of radiant heat.

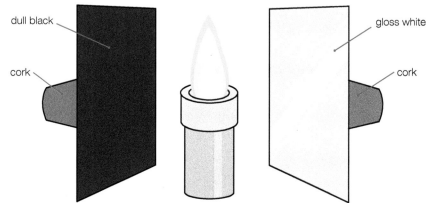

dull black

cork

gloss white

cork

▲ **Figure 6.11** Investigating radiation absorption

Rules to remember:

▶ Radiant heat falling on a surface is partly absorbed and partly reflected.

▶ Matt black surfaces are good absorbers (and poor reflectors) of radiation.

▶ Gloss white surfaces are poor absorbers (and good reflectors) of radiation.

Radiation summary

Table 6.2

Matt black surfaces are	Gloss white surfaces are
Good emitters and absorbers of radiation	Poor emitters and absorbers of radiation
Poor reflectors of radiation	Good reflectors of radiation

Global warming

Most of the infrared radiation from the Sun that arrives at the Earth is reflected back into space, but some of it reaches the Earth's surface where it is absorbed (Figure 6.12). The Earth itself then radiates part of this energy back into its atmosphere.

The infrared energy that is radiated by the Earth has a longer wavelength than the waves that come from the Sun. This long-wavelength radiation is then absorbed by the carbon dioxide (and other greenhouse gases such as methane and water vapour) in the atmosphere, keeping the Earth warm. This effect is known as the normal greenhouse effect and without it, life on Earth would be impossible for humans.

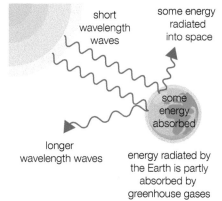

short wavelength waves

some energy radiated into space

some energy absorbed

longer wavelength waves

energy radiated by the Earth is partly absorbed by greenhouse gases

▲ **Figure 6.12** The greenhouse effect

As more and more fossil fuels are burned, the amount of carbon dioxide in the atmosphere is increasing. This has the effect of absorbing more of the energy radiated from the Earth, which makes the atmosphere warmer. This effect is called **global warming**.

Some consequences of increasing global temperatures are:

▶ polar ice-caps begin to melt, causing widespread flooding

▶ the occurrence of extreme weather events (such as Hurricane Katrina, which devastated New Orleans) that can lead to droughts in some countries and floods in others

▶ a reduction in agricultural yields, leading to famine in some countries

▶ destruction of some habitats, leading to species extinctions

▶ increases in the spread of disease.

Many scientists are concerned that global warming might increase so much that it becomes irreversible. As the air temperature increases, more and more dissolved carbon dioxide passes from oceans, lakes and rivers into the atmosphere. This causes more global warming, which causes further carbon dioxide to pass into our atmosphere, and so the process continues. This effect is called the runaway greenhouse effect, and would be catastrophic for life on Earth.

There is ongoing political and public debate worldwide regarding what action, if any, should be taken to reduce greenhouse gas emissions. Developed countries will certainly be required to burn less fossil fuels, and so reduce their emissions of carbon dioxide. This will be difficult as people adjust to relying less on oil-powered transport and to burning less fossil fuels. There will be intense debate about the increasing use of nuclear power and renewable resources.

Applications of heat transfer

Vacuum flasks

The vacuum flask was designed by James Dewar in order to keep liquids cold. But the flask works equally well as a way of keeping liquids hot. Today, it is commonly used as a picnic flask to keep tea, coffee or soup hot. How does it work? Look at Figure 6.13.

▶ The flask is made of a double-walled glass bottle. There is a vacuum between the two walls. The vacuum stops heat transfer by conduction or convection through the sides.

▶ The glass walls facing the vacuum are silvered. Their shiny surfaces reduce heat transfer by radiation to a minimum.

▶ The stopper is made of plastic, and is often filled with cork or foam to reduce heat transfer by conduction through it.

▶ The outer, sponge-packed plastic case protects the inner, fragile flask against physical damage.

Show you can ❓

a) Give a brief account of the greenhouse effect, and explain how it leads to global warming.

b) Give one harmful and one beneficial effect global warming might have for the people of Ireland.

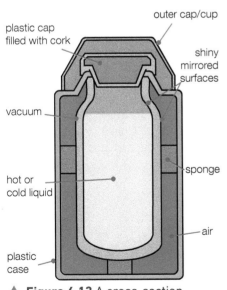

plastic cap filled with cork

outer cap/cup

shiny mirrored surfaces

vacuum

hot or cold liquid

sponge

air

plastic case

▲ **Figure 6.13** A cross-section through a vacuum flask

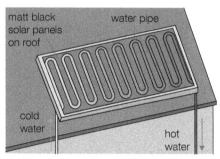

▲ **Figure 6.14** How solar water heating works

Solar water heating

A solar hot water panel absorbs sunlight and uses the energy to heat water (Figure 6.14) in the following process:

▶ The sunlight passes through a glass window and falls onto a blackened metal sheet.

▶ The metal is in a draught-proof enclosure to minimise heat loss by convection.

▶ The blackened metal absorbs almost all of the energy in the sunlight and its temperature often rises to over 100 °C.

▶ The heat stored in the blackened metal is then transferred to water flowing in the attached pipe.

Solar water heating is ideal for use where large volumes of hot water are needed, such as swimming pools and hospital laundries.

Test yourself

1 As food is cooled in a fridge, heat energy is transferred to a coolant. The coolant (usually a liquid with a low boiling point) passes through pipes at the back of the fridge (Figure 6.15). The pipes are usually painted black and have thin metal 'fins' attached.

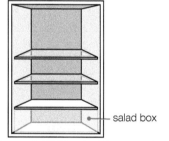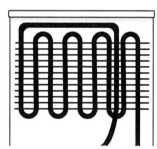

▲ **Figure 6.15** The inside of a fridge

  a) Why are the pipes painted black?
  b) Why are the pipes mounted on thin metal fins?

2 Which part of an oven is hottest, the top or the bottom? Why is this so? What is the purpose of the fan in a fan-assisted oven?

3 Suggest a reason why the roofs of houses in hot countries are often painted white.

4 Computer chips can produce a lot of heat. If the chips become too hot they might get damaged. They are often attached to a heavy piece of copper metal with black fins, as shown in Figure 6.16. Such a piece of metal is called a heat sink. What four features of the design of this type of heat sink make it suitable for its purpose?

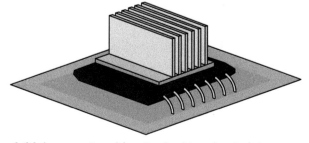

▲ **Figure 6.16** A computer chip attached to a heat sink

1 a) Copy and complete the table below to indicate the main method of heat transfer in each substance.

Substance	Method of heat transfer
Water	
Copper	
Glass	

(3 marks)

Water is heated in a glass saucepan (Figure 6.17).

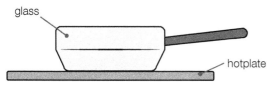

Figure 6.17

b) Describe, in terms of particles, how heat is transferred through the glass from the hotplate to the water. *(2 marks)*

2 Two rods, one made of glass and the other of iron, are placed in a Bunsen flame (Figure 6.18). The dimensions of the rods are exactly the same.

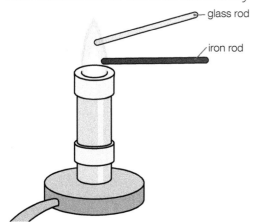

Figure 6.18

Copy and complete the table below. After each of the statements, write the letter G if the statement applies to glass. Write I if the statement applies to iron. If the statement applies to both rods, then put GI in the box.

This rod has no free electrons.	
Atoms are mainly responsible for heat conduction.	
Atoms vibrate more quickly when heat is added.	
Heat is transferred when electrons collide with neighbouring atoms.	

(2 marks)

3 Jenny heats a beaker of water (Figure 6.19).

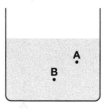

Figure 6.19

a) Draw arrows at point A and B to show the direction of water movement. *(2 marks)*

Two conical flasks contain the same amount of water at the same temperature. They are placed at equal distances from a radiant heater (Figure 6.20).

Figure 6.20

b) Explain why the temperature of the water in the flask with the dull black paint rises more quickly than in the flask with the shiny silver paint.

(2 marks)

4 The apparatus shown in Figure 6.21 was set up to study how heat was transmitted through different materials. After a few minutes of heating, the wax on the mercury-filled test tube melts and the cork falls off. The wax on the water-filled test tube takes much longer to melt.

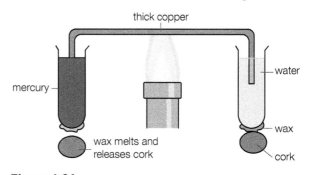

Figure 6.21

a) Name the main method by which heat from the Bunsen flame reaches the mercury and the water. *(1 mark)*

b) What does this experiment tell you about heat transfer in mercury and water? *(2 marks)*

Figure 6.22 shows a simple experiment to examine the heat given out by a Bunsen burner. The corks are attached to the metal plates by wax, which soon melts.

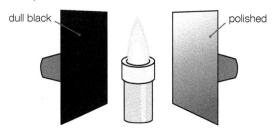

dull black — polished

Figure 6.22

c) Name the main method by which heat reaches the metal plates from the Bunsen flame. *(1 mark)*

d) Describe and explain what happens after a few minutes of heating. *(2 marks)*

5 a) An electric kettle is half full of water (Figure 6.23).

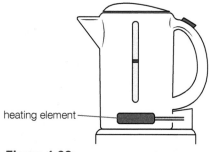

heating element

Figure 6.23

i) Name the process by which heat energy is spread throughout the water. *(1 mark)*

ii) Copy the diagram of the kettle and use arrows to draw the directions of the currents in the water generated by the heating element. *(1 mark)*

b) A saucepan containing cold water is placed on a hotplate. The water is heated through the base of the saucepan. Name this process of heat transfer. *(1 mark)*

c) The base of the saucepan is made of steel, and the handle is made of wood. These materials are chosen for their thermal properties.

i) Why is steel used for the base? *(1 mark)*

ii) Why is wood used for the handle? *(1 mark)*

6 Figure 6.24 shows two spoons being held in ice water. One is made of steel, and the other is made of plastic.

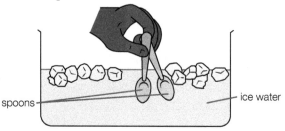

spoons — ice water

Figure 6.24

a) Which one will feel colder? *(1 mark)*

b) Explain the reason for your answer to **(a)**. *(1 mark)*

7 Figure 6.25 below shows an experiment in which water can be boiled at the top of a test tube while ice cubes are held at the bottom of the tube by a metal gauze.

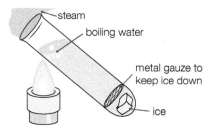

steam

boiling water

metal gauze to keep ice down

ice

Figure 6.25

a) i) Why do convection currents not carry heat down to the ice? *(1 mark)*

ii) What does this experiment tell us about the ability of water to conduct heat? *(1 mark)*

Figure 6.26 shows an experiment sometimes used in radiation investigations. The four side surfaces of the box are equal in size and are painted. One is matt black, another is shiny black, another is shiny white and the fourth is matt white. The metal box is filled with very hot water. Identical thermometers are held equal distances from the four side surfaces of the metal box, and the temperature on each is recorded.

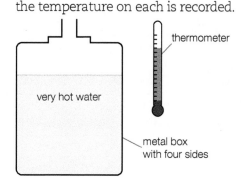

thermometer

very hot water

metal box with four sides

Figure 6.26

b) i) Explain why the experiment works best if the thermometer bulbs are blackened. *(1 mark)*

ii) Why is it important that the thermometers are all the same distance from the surface of the box? *(1 mark)*

iii) Which one of the four surfaces gives out the most radiant heat? *(1 mark)*

iv) How could you tell from the thermometers which surface was giving out most radiant heat? *(1 mark)*

a) i) Copy the diagram. Draw arrows at points X and Y to show the direction of water flow. *(2 marks)*

ii) Name the main process by which heat travels through the water. *(1 mark)*

b) One method of heat transfer can take place in a vacuum. What is the name of this method of heat transfer? *(1 mark)*

8 Purple dye crystals may be used to trace heat transfer in water that is being heated with a Bunsen burner (Figure 6.27).

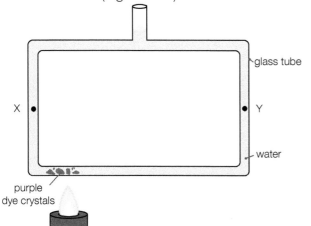

purple dye crystals

glass tube

X

Y

water

Figure 6.27

7 Atomic and nuclear physics

Specification points

This chapter covers sections 1.5.1 to 1.5.27 of the GCSE Physics specification (and 1.5.1 to 1.5.23 of the Double Award specification). Note: The section about ITER on page 93 is **not** in the Double Award specification. You will explore the particle structure of both the atom and the nucleus, and examine radioactivity as a consequence of unstable nuclei and study the properties of alpha, beta and gamma radiation. You are introduced to the terms background and half-life and discuss the damaging effect that nuclear radiations have on our bodies. You also learn about fusion and fission as sources of energy.

The structure of atoms

We take it for granted today that all matter is made up of atoms, but what are atoms made of? Experiments carried out by J.J. Thomson and Ernest Rutherford led physicists in the early part of the twentieth century to believe that atoms themselves had a structure.

Evidence for the existence of electrons

When a current passes through a metal wire, the wire gets hot. If the wire is hot enough, it emits negatively charged particles. If the wire itself is connected to the negative terminal of a battery, these negatively charged particles are repelled from the wire, which is called the **cathode**. They can be collected by a positive plate, which is called the **anode**. The wire and plate must be set in a vacuum if the negatively-charged particles are not to be deflected by collisions with gas atoms (see Figure 7.1).

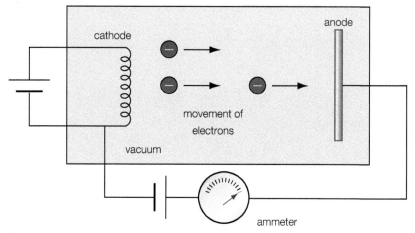

▲ **Figure 7.1** A current being passed through a metal wire

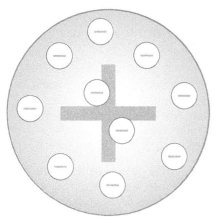

▲ **Figure 7.2** The 'plum pudding' model of the atom

The process is called **thermionic emission**. Early in the twentieth century, physicists were able to show that these particles came from the atoms of the metal filament. Thomson called them **electrons**. Their attraction to the positive plate showed clearly that they were negatively charged.

Since these electrons were so easily deflected by a magnet, Thomson knew that they were very, very light compared with the atoms that emitted them. He also realised that since atoms themselves are electrically neutral, there had to be some part of the atom that had a positive charge.

One of the earliest models of the atom was the 'plum pudding' model (Figure 7.2), in which electrons were dotted throughout the atom like currants in a bun. The positive charge was thought to spread throughout the volume like the dough of the bun.

Rutherford's nuclear model

In 1911, partly to test Thomson's theory, Rutherford devised an experiment in which recently discovered positively charged alpha particles were fired at a thin gold foil (Figure 7.3). The observations from the experiment were:

▶ Most of the alpha particles went straight through the foil.

▶ A few alpha particles were slightly deflected.

▶ Some were deflected through very large angles.

▶ 1 in 8000 were 'back scattered' (reflected back in the direction from which they came).

These observations could not be explained by the 'plum pudding' model.

What really shocked Rutherford's team was that some alpha particles were deflected through very large angles, and a few even came straight back at them. Rutherford then realised that there had to be something 'hard' inside the atom to cause this strange 'back scattering'. He called it the atomic nucleus.

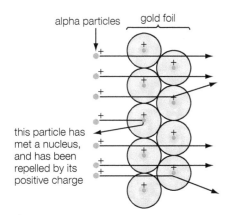

alpha particles   gold foil

this particle has met a nucleus, and has been repelled by its positive charge

▲ **Figure 7.3** Most of the alpha particles pass straight through the gold foil or are slightly deflected. A very small number make a 'direct hit' on the nucleus and bounce back

Rutherford's deductions:

1 The majority of the alpha particles passed straight through the metal foil because they did not come close enough to any repulsive positive charge at all. The atom must be mostly empty space.

2 All the positive charge and most of the mass of an atom formed a dense core (the nucleus).

3 The negative charge consisted of a 'cloud of electrons' surrounding the positive nucleus.

4 Only when a positive alpha particle approached sufficiently close to a nucleus was it repelled strongly enough to rebound at large angles.

5 The small number of alpha particles that were back scattered indicated that the nucleus was small and that it contained nearly all the mass of the atom.

In 1913, in order to explain how certain elements gave out light, Niels Bohr suggested that the electrons orbited the nucleus in circular paths. This model is called the **Rutherford–Bohr model**, in which the orbiting electrons are pictured like planets orbiting the Sun. This model has been developed further to more fully explain the behaviour of electrons within atoms.

Rutherford's gold foil experiment took place around 1909–10. It was not until 1933 that convincing evidence was presented by James Chadwick that there were two different types of particle in the nucleus – uncharged neutrons as well as positively charged protons.

The structure of the nucleus

An atom is made up of smaller particles. There is a central nucleus made up of protons and neutrons. Around this, **electrons** orbit at high speed. The number of particles depends on the type of atom. Protons have a positive (+) charge. Electrons have an equal negative (−) charge. Normally, atoms are neutral. So an atom must have the same number of electrons as protons. Protons and neutrons are collectively called **nucleons**. Each is about 1800 times more massive than an electron, so virtually all of an atom's mass is in its nucleus. Electrons are held in orbit by the force of attraction between opposite charges.

The relative masses and charges of the particles that make up the atom are given in Table 7.1.

Table 7.1

Particle	Location	Relative mass*	Relative charge*
Proton	Within the nucleus	1	+1
Neutron	Within the nucleus	1	0
Electron	Orbiting the nucleus	$\dfrac{1}{1840}$	−1

*Mass and charge are measured relative to the proton

Figure 7.4 shows a simple representation of a helium atom. Since there are two orbiting electrons, there must also be two protons in the nucleus. Note that the diagram is not to scale: the diameter of the atom (about 1×10^{-10} m) is about 100 000 times greater than that of the nucleus (about 1×10^{-15} m).

Atomic number and mass number

The number of protons in the nucleus of an atom determines what type the atom is. All hydrogen nuclei have one proton, all helium nuclei have two protons, and all lithium nuclei have three protons, and so on. The number of protons is called the **atomic number** and is given the symbol **Z**. The atomic number also tells you the number of electrons in the atom.

As the mass of the electrons is negligible, the total number of particles in the nucleus determines the total mass of an atom. The **mass number** (or nucleon number) is the sum of the number of protons and the number of neutrons. The mass number is given the symbol **A**.

atomic number, **Z** = number of protons

mass number, **A** = number of protons + number of neutrons
　　　　　　　 = number of nucleons

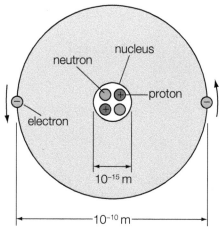

▲ **Figure 7.4** A helium atom

Every nucleus can therefore be written in the form: $^A_Z X$

where **X** is the chemical symbol, A is the mass number and Z is the atomic number.

For example, the element uranium has the chemical symbol U. All uranium nuclei have 92 protons in the nucleus. One form of uranium, called uranium-235, has a mass number of 235. This means it has 92 protons and 143 neutrons (235 – 92 = 143). A uranium nucleus is given the symbol:

mass number —— 235
atomic number —— $_{92}$ U —— element symbol

It is important to realise that this is the symbol for the nucleus of the atom. Orbiting electrons are completely ignored.

You will notice that the top number gives the mass of the nucleus, and the bottom number gives the charge. This same system can also be used to describe protons, neutrons and electrons.

proton $^1_1 p$ neutron $^1_0 n$ electron $^0_{-1} e$

Isotopes

Not all the atoms of the same element have the same mass. For example, one form of helium (helium-3) has three nucleons, and another form (helium-4) has four nucleons (Figure 7.5). But all helium nuclei have two protons. So, helium-3 has two protons and one neutron, helium-4 has two protons and two neutrons.

Atoms with the same number of protons but a different number of neutrons are called isotopes.

Isotopes are atoms of the same element that have the same atomic number but different mass numbers.

The main isotopes of helium are $^3_2 He$ and $^4_2 He$.

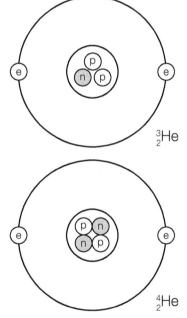

▲ **Figure 7.5** The structures of helium-3 and helium-4

$^3_2 He$

$^4_2 He$

Test yourself

1   An atom contains electrons, protons and neutrons. Which of these particles:
   a) are outside the nucleus?
   b) are uncharged?
   c) have a negative charge?
   d) are nucleons?
   e) are much lighter than the others?
2   How many protons, neutrons and electrons are in the nucleus of carbon-14 if its symbol is $^{14}_6 C$?
3   The element sodium has the chemical symbol Na. In a particular sodium isotope, there are 12 neutrons. In a neutral sodium atom, there are 11 orbiting electrons. Write down the symbol for the nucleus of this isotope.
4   In what way are the nuclei of isotopes the same? In what way are they different?

Nuclear radiation

In 1896, the French scientist Henri Becquerel discovered by accident that certain rocks containing uranium gave out strange radiation that could penetrate paper and affect photographic film. He called the effect radioactivity. His students, Pierre and Marie Curie, later went on to identify three separate types of radiation. Unsure of a suitable name, the Curies called them alpha (α), beta (β) and gamma (γ) radiation after the first three letters of the Greek alphabet. The Curies and Henri Becquerel were jointly awarded the Nobel Prize for Physics in 1903 for their work on radioactivity.

For very heavy elements such as uranium or plutonium, the large number of protons and neutrons can make the nucleus unstable and cause their nuclei to randomly and spontaneously emit radiation. The atoms that emit such radiation are said to be radioactive. The particles and waves are referred to as nuclear radiation. The materials are called radioactive materials. The disintegration is called radioactive decay.

Ionising radiation

Ions are charged atoms (or molecules). Atoms become ions when they lose (or gain) electrons.

Nuclear radiation can become dangerous by removing electrons from atoms in its path, so it has an ionising effect (see Figure 7.6).

When this happens with molecules of genes in living cells, the genetic material of a cell is damaged and there is a small chance that the cell may become cancerous. Other forms of ionising radiation include ultraviolet and X-rays.

Nature and properties of nuclear radiations

The three main types of nuclear radiation are:

Alpha radiation α or $^{4}_{2}$He

▶ Alpha radiation is made up of a stream of alpha particles emitted from large nuclei.

▶ An alpha particle is a helium nucleus with two protons and two neutrons, and so has relative atomic mass of 4.

▶ Alpha particles are positively charged and so will be deflected in a magnetic field.

▶ Alpha particles have poor powers of penetration and can only travel through a few centimetres of air. They can easily be stopped by a sheet of paper.

▶ Alpha radiation has the strongest ionising power.

▶ Alpha radiation is not as dangerous if the radioactive source is outside the body, because it cannot pass through the skin and is unlikely to reach cells inside the body.

▶ Alpha radiation will damage cells if the radioactive source has been breathed in or swallowed.

An example of decay by alpha emission is shown in Figure 7.7.

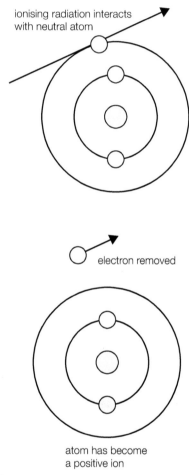

ionising radiation interacts with neutral atom

electron removed

atom has become a positive ion

▲ **Figure 7.6** Ionisation caused when an electron is removed from an atom by radiation

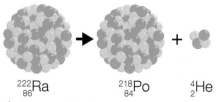

$^{222}_{86}$Ra $^{218}_{84}$Po $^{4}_{2}$He

▲ **Figure 7.7** Alpha decay

Beta radiation β or $_{-1}^{0}e$

Beta radiation is made up of a stream of beta particles emitted from nuclei where the number of neutrons is much larger than the number of protons.

▶ A beta particle is a fast electron which has been formed in the nucleus and so has relative atomic mass of about $\dfrac{1}{1840}$.

▶ As beta particles are negatively charged, they will be deflected in a magnetic field. This deflection will be greater than that of alpha particles as beta particles have a much smaller mass.

▶ Beta particles move much faster than alpha particles, and so have greater penetrating power.

▶ Beta particles can travel several metres in air, but will be stopped by 5 mm thick aluminium foil.

▶ Beta radiation has an ionising power between that of alpha and gamma radiation.

▶ Beta radiation can penetrate the skin and cause damage to cells.

An example of beta decay is shown in Figure 7.8.

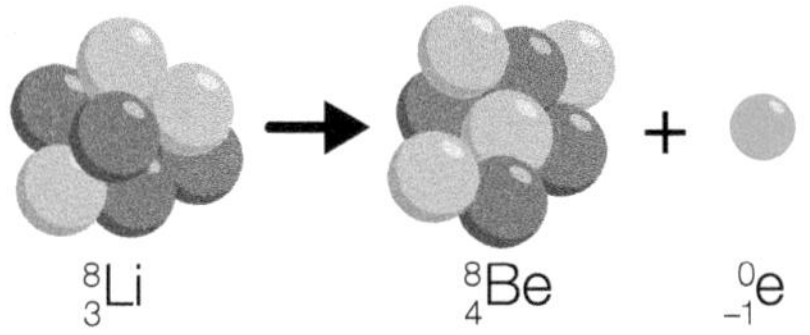

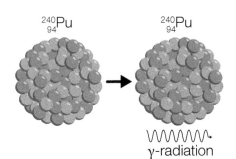

▲ **Figure 7.8** Beta decay

Gamma radiation γ

▶ Unlike the other types of radiation, gamma radiation does not consist of particles but of high-energy waves.

▶ Like alpha and beta radiation, gamma radiation comes from a disintegrating unstable nucleus. As it is an electromagnetic wave (see Chapter 8), gamma radiation has no mass.

▶ As there are no charged particles, a magnetic field has no effect on gamma radiation.

▶ Gamma radiation has great penetrating power, travelling several metres through air.

▶ A thick block of lead or concrete is used to greatly reduce the effects of gamma radiation, but is not able to stop it completely.

▶ Gamma radiation has the weakest ionising power.

▶ Gamma radiation can penetrate the skin and cause damage to cells.

An example of gamma emission is shown in Figure 7.9.

The penetrating power of each of these emissions is summarised in Figure 7.10.

▲ **Figure 7.9** Gamma decay

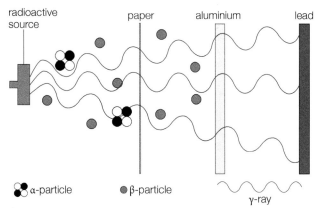

▲ **Figure 7.10** Selective absorption of radioactive emissions

5 Name a radioactive isotope that occurs naturally in living things.
6 Copy and complete the table below.

Property	Alpha particles (α)	Beta particles (β)	Gamma rays (γ)
Nature	each particle is made up of two protons and two neutrons, so it is identical to a nucleus of helium		very high energy electromagnetic waves
Relative charge compared with charge of a proton		−1	0
Mass	high compared to beta particles	low	0
Speed	up to 0.1 × speed of light		speed of light
Ionising effect	strong	weak	very weak
Penetrating effect		penetrating, but stopped by a few millimetres of aluminium or other metal	very penetrating: never completely stopped, though lead and thick concrete will reduce intensity

Which of the three types of radiation:

a) is a form of electromagnetic radiation?
b) carries positive charge?
c) is made up of electrons?
d) travels at the speed of light?
e) is the most ionising?
f) can penetrate a thick sheet of lead?
g) is stopped by skin or thick paper?

Dangers of radiation

Most radioactive background activity comes from natural sources such as cosmic rays from space or from rocks and soil, some of which contain radioactive elements such as radon gas (see Figure 7.11). Living things and plants absorb radioactive materials from the soil, which are then passed along the food chain. There is little we can do about natural background radiation, although people who live in areas with a high background radiation level due to radon gas require homes to be well ventilated to remove the gas. Human behaviour also adds to the background activity that we are exposed to through medical X-rays, radioactive waste from nuclear power plants and the radioactive fallout from nuclear weapons testing.

Radioactive material is found naturally all around us and inside our bodies. A small number of carbon atoms occurring naturally are radioactive carbon-14 isotopes. They can be found in the carbon

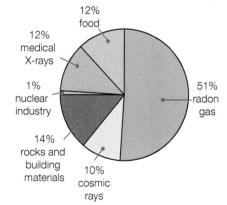

▲ **Figure 7.11** Sources of background radiation

dioxide in the air and in the cells of all living organisms. Traces of radioactive elements, such as potassium, can be found in our food. Certain rocks contain uranium (all the isotopes of which are radioactive) and this decays into radon, a radioactive gas. There is also radiation reaching Earth from outer space, referred to as cosmic rays. All these natural sources are known together as background radiation.

Protection when handling

You can minimise the risk to those using radioactive materials by:

▶ wearing protective clothing
▶ keeping the source as far away as possible by using tongs
▶ limiting exposure time to as little as possible
▶ keeping radioactive materials in lead-lined containers.

Nuclear disintegration equations

Symbol equations can be written to represent alpha and beta decay.

The alpha particle can be written as $^4_2$He and the beta particle as $^0_{-1}$e.

Examples

1 Alpha decay of uranium-238

$$^{235}_{92}U \rightarrow {}^{234}_{90}Th + {}^4_2He$$

2 Beta decay of carbon-14

$$^{14}_{6}C \rightarrow {}^{14}_{7}N + {}^0_{-1}e$$

When writing symbol equations, it is important to remember the following:

▶ The sum of the mass numbers on the left-hand side of the equation must equal the sum of the mass numbers on the right-hand side.
▶ The sum of the atomic numbers on the left-hand side of the equation must equal the sum of the atomic numbers on the right-hand side.

If you know the original isotope (referred to as the parent nucleus) and the isotope that was formed by the decay (known as the daughter nucleus), it is possible to determine the type of decay by working out the type of particle emitted.

Or, if you know the original isotope and the type of decay, you can work out the isotope that is formed by the decay.

As you can see in the worked example, alpha decay results in the mass number of the parent nucleus decreasing by four and its atomic number decreasing by two.

Alpha decay is exemplified by the equation below.

$$^A_ZX \rightarrow {}^{A-4}_{Z-2}Y + {}^4_2He$$

For example, uranium-235 decays by emission of an α-particle into thorium-231:

$$^{235}_{92}U \rightarrow {}^{231}_{90}Th + {}^4_2He$$

Example

Radium-226 decays to polonium-222. Radium (Ra) has atomic number 86 and polonium (Po) has atomic number 84. What is 'X'? Which type of decay occurs?

$$^{226}_{86}Ra \rightarrow {}^{222}_{84}Po + {}^a_bX$$

Answer

Balancing mass numbers:

226 = 222 + a

a = 4

Balancing atomic numbers:

86 = 84 + b

b = 2

So X is a helium nucleus (alpha particle).

This must be alpha decay.

However, if the mass number of the parent does not change and the atomic number of the daughter nucleus increases by 1, then the reaction must be beta decay.

Beta decay is exemplified by:

$$_Z^A X^* \rightarrow _{Z+1}^A Y + _{-1}^0 e \ (\text{or } _{-1}^0 \beta)$$

To take a specific case, radium-228 decays by emitting a β-particle to form actinium-228:

$$_{88}^{228}Ra \rightarrow _{89}^{228}Ac + _{-1}^0 e$$

In gamma decay, the parent nucleus de-excites, and emits gamma ray(s) in the process. There is no change in the nature of the nucleus, so the mass number and the atomic number stay the same.

The γ-radiation is usually emitted at the same time as the α- and β-particle emissions, and represents the excess energy of the daughter nucleus as it settles down into a more stable condition.

$$_Z^A X \rightarrow _Z^A X + \gamma$$

Test yourself

7 Copy and complete the table below.

Radiation	Atomic number (Z)	Mass number (A)
α-emission	decreases by 2	
β-emission		unchanged
γ-emission		

8 Copy and complete the following equations for alpha decay:

a) $_{92}^{238}U \rightarrow \boxed{} Th + _2^4 He$

b) $\boxed{}_{92} Pu \rightarrow _{92}^{238}U + _2^4 He$

c) $_{\boxed{}}^{251}Cf \rightarrow _{96}^{\boxed{}} Cm + _2^4 He$

d) $\boxed{} Hf \rightarrow _{70}^{170} Yb + _2^4 He$

e) $\boxed{} Bi \rightarrow _{81}^{207} Tl + _2^4 He$

f) $_{\boxed{}}^{190}Pt \rightarrow _{76}^{\boxed{}} Os + _2^4 He$

9 Copy and complete the following equations for beta decay:

a) $_6^{14}C \rightarrow \boxed{} N + _{-1}^0 e$

b) $\boxed{}_1 H \rightarrow _{\boxed{}}^3 He + _{-1}^0 e$

c) $_{\boxed{}}^{137}Cs \rightarrow _{56}^{\boxed{}} Ba + _{-1}^0 e$

d) $\boxed{}_{19} K \rightarrow _{\boxed{}}^{40} Ca + _{-1}^0 e$

e) $\boxed{} Co \rightarrow _{28}^{60} Ni + _{-1}^0 e$

f) $_{15}^{32}P \rightarrow \boxed{} S + _{-1}^0 e$

10 Work out the type of decay in each of the following examples:
   a) bismuth-213 to polonium-213
   b) radium-226 to radon-222
   c) francium-221 to actinium-217
11 a) How does the value of the mass number change in alpha decay?
   b) How does the value of the atomic number change in alpha decay?
   c) How does the value of the mass number change in beta decay?
   d) How does the value of the atomic number change in beta decay?

Radioactive decay

For very heavy elements such as uranium or plutonium, the large number of protons and neutrons can make the nucleus unstable and cause these nuclei to undergo radioactive decay in a **random** and **spontaneous** manner. Random means that we cannot predict when a particular nucleus will disintegrate. Spontaneous means that the rate of decay is unaffected by any physical changes such as temperature, pressure or chemical changes. Some types of nuclei are more unstable than others and decay at a faster rate.

Rate of decay and half-life

As a radioactive isotope decays, the activity of the sample decreases. This is because as the atoms of the original sample disintegrate, there will gradually be fewer and fewer original atoms left to disintegrate. To illustrate this process, consider radioactive iodine-131, which has a half-life of eight days. Imagine there are 40 billion atoms present at an instant in time. Look at Figure 7.12. In this diagram, one dot represents 1 billion atoms.

After eight days, 20 billion atoms will have decayed, leaving only 20 billion radioactive atoms. After another eight days, 10 billion more would have decayed, leaving only 10 billion radioactive atoms. After a further eight days, 5 billion more would have decayed, leaving only 5 billion radioactive atoms, and so on. As a result, the activity (the number of particles decaying in a particular time) decreases.

The half-life of a radioactive substance is the time taken for half the nuclei in any sample of the substance to decay. It follows that half-life of a substance is the time taken for the count rate to fall to half its original value.

The half-life of a radioactive isotope is defined as the time taken for its activity to fall by half.

Each isotope has a specific and constant half-life. Some half-lives are very short – a matter of seconds or even a fraction of a second – and others can be thousands of years. Figure 7.13 shows a graph of the radioactive activity of a sample of an isotope with a half-life of 2 hours. Table 7.2 gives the half-lives of some common radioactive isotopes.

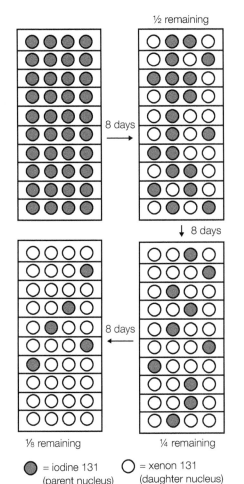

½ remaining

8 days

↓ 8 days

8 days

⅛ remaining   ¼ remaining

● = iodine 131 (parent nucleus)   ○ = xenon 131 (daughter nucleus)

▲ **Figure 7.12** Illustration of half-life

Table 7.2

Isotope	Half-life
uranium-238	4 500 000 000 years
carbon-14	5730 years
phosphorous-30	2.5 minutes
oxygen-15	2.06 minutes
barium-144	114 seconds
polonium-216	0.145 seconds

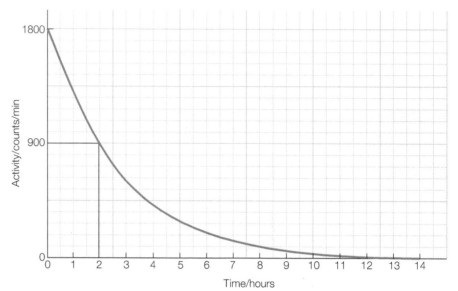

▲ **Figure 7.13** The radioactive decay curve for a substance with a half-life of 2 hours

The unit of radioactivity

The unit for radioactivity is the Becquerel (Bq).

1 Bq = 1 disintegration per second.

So if a radioactive material emits 1800 alpha particles every minute and 30 nuclei decay every second, its activity is 30 Bq.

When doing calculations, it can be very helpful to know the following:

▶ After one half-life, half the radioactive material remains.

▶ After two half-lives, a quarter of the radioactive material remains.

▶ After three half-lives, an eighth of the radioactive material remains.

▶ After four half-lives, a sixteenth of the radioactive material remains, and so on.

Example

1 What mass of nitrogen-13 would remain if 80 g were allowed to decay for 30 minutes? Nitrogen-13 has a half-life of 10 minutes.

Answer

Calculations like this can be easily done using a table. Take note of the headings in the table.

Mass of nitrogen-13 remaining	Time/half-lives	Time/minutes
80 g	0	0
40 g	1	10
20 g	2	20
10 g	3	30

So 10 g would remain after 30 minutes.

2 Strontium-93 takes 32 minutes to decay to 6.25% of its original mass. Calculate the value of its half-life.

Answer

From the table below, four half-lives take 32 minutes.

each half-life = 32 ÷ 4 minutes

= 8 minutes

% of strontium-93 remaining	Time/half-lives	Time/minutes
100	0	0
50	1	8
25	2	16
12.5	3	24
6.25	4	32

Test yourself

12 Calculate the half-lives of the following samples:
   a) A sample of iodine-123 whose activity falls from 1000 Bq to 250 Bq in 14.4 hours.
   b) A sample of technetium-99 whose activity falls from 200 Bq to 25 Bq in 18 hours.
   c) A sample of strontium-90 whose activity falls from 500 Bq to 62.5 Bq in 86.4 years.

13 Calculate how long it would take for the following to decay to an activity of 1 Bq:
   a) A sample of cobalt-60 (half-life = 5.27 years) whose original activity is 64 Bq.
   b) A sample of iodine-131 (half-life = 8 days) whose original activity is 128 Bq.
   c) A sample of polonium-210 (half-life = 138 days) whose original activity is 32 Bq.

14 How long would it take for 20 g of cobalt-60 to decay to 5 g? The half-life of cobalt-60 is 5.26 years.

15 When a radioactive material with a half-life of 24 hours arrives in a hospital, its activity is 1000 Bq. Calculate its activity 24 hours before and 72 hours after its arrival. (Hint: Draw up a table as shown below.)

Activity in Bq	Time/half-lives	Time/hours
	−1	−24
1000 (start from here)	0	0
500	1	24

16 Plot a graph of activity (y-axis) against time (x-axis) using the data in question 15. Start the graph from time = 0 and activity = 1000 Bq. Use the graph to find the activity 36 hours after the material arrives at the hospital.

Uses of radiation

In medicine

Gamma radiation from the cobalt-60 isotope can be used to treat tumours (Figure 7.16).

Different radioisotopes are used to monitor the function of organs by injecting a small amount into the bloodstream and detecting the emitted radiation. The tracers used in this case must have a short half-life.

Iodine-131 is used in investigations of the thyroid gland (Figure 7.14).

Surgical instruments and hospital dressings can be sterilised by exposure to gamma radiation (Figure 7.15). The source should have a very long half-life so that it does not need to be replaced on a regular basis.

Great care must be taken when using radioactive isotopes because the radiation can damage living cells by altering the structure of the cell's chemicals. Protective clothing must be worn, and the amount of time that the worker is exposed to the radiation must be strictly controlled. Radioactive isotopes that are taken internally are usually not alpha emitters (as they are such powerful ionisers), and they must have a short half-life so that they do not remain for too long in body tissues.

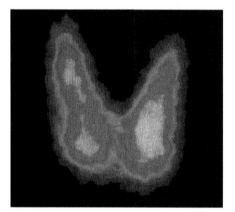

▲ **Figure 7.14** This scan shows radioactive iodine-131 localised in the thyroid gland

▲ **Figure 7.15** Sterilising equipment using gamma rays

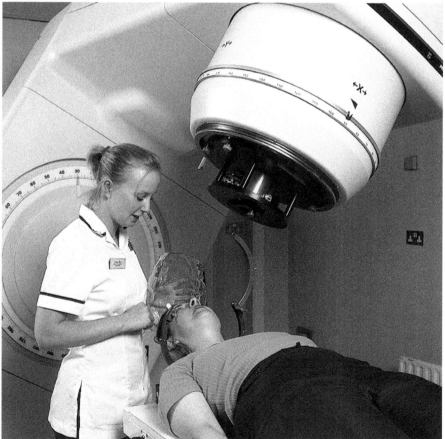

▲ **Figure 7.16** Radiotherapy involves the use of radioactive materials to treat cancers

In agriculture

Gamma radiation can be used to treat fresh food (Figure 7.17a). By killing bacteria on the food, the radiation helps the food to have a longer shelf-life. It is important to remember that this irradiation does not make the food itself radioactive. The use is controversial, however, as many people are worried about the long-term effects on the human body of eating irradiated food. Ideally, the radioisotopes used in food processing plants should have a very long half-life so that it is a long time before they need to be replaced.

The ease with which a plant absorbs a fertiliser can be found by putting a small amount of radioactive isotope in the fertiliser (Figure 7.17b). You can tell how much fertiliser has been taken up by the plant by checking different parts of the plant for radioactivity.

a)   b)

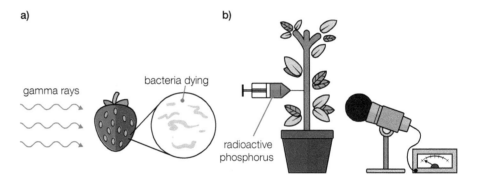

▲ **Figure 7.17** a) Sterilising food and b) measuring fertiliser take-up by a plant

In industry

Beta radiation can be used to monitor the thickness of a sheet of paper or aluminium (Figure 7.18). An emitter is placed on one side of the sheet and a detector on the other. As the sheet moves past, the activity detected will be the same as long as the thickness remains unchanged.

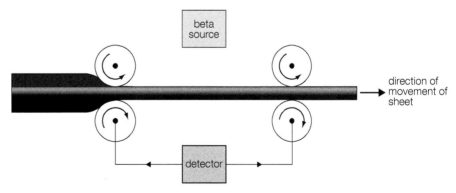

▲ **Figure 7.18** A long half-life beta source is used to control the thickness of an aluminium sheet

Radioactive tracers

A suitable radioactive isotope can be used to provide information about fluid movement and mixing to monitor things like leaks in underground pipes (Figure 7.19). The radiation needs to penetrate many centimetres of soil to reach the detectors, which means that it must be a gamma emitter, because this is the only type of radiation with sufficient penetrating power. To avoid dangerous radioactive materials being in the ground for a long time, the source should have a short half-life.

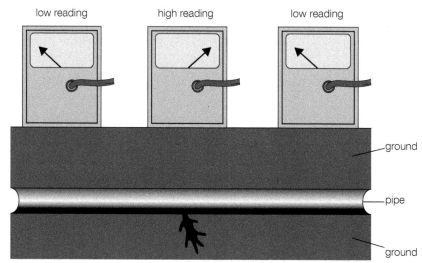

▲ **Figure 7.19** Radioactive tracers can be used to locate a leak in a pipe

In the home

In an ionisation smoke detector (Figure 7.20), a source of alpha radiation (often americium-241) is placed in the detector close to two electrodes. Ions are formed in the air around the radioactive source, and these allow a tiny current to flow.

If there is a fire, smoke will block the path of the ions, causing the current to fall. The fall in current is detected electronically, and a siren is sounded.

Practical work with radioactive materials

Students under the age of 16 are expressly forbidden from handling radioactive sources. However, you should know how practical work can be carried out.

The most common type of radiation detector is the Geiger–Müller tube (GM tube) connected to a counter (Figure 7.21).

When alpha, beta or gamma radiation enters the GM tube, it causes some of the argon gas inside to ionise and give an electrical discharge. This discharge is detected and counted by the counter. If the counter is connected to its internal speaker, you can hear the click when radiation enters the tube.

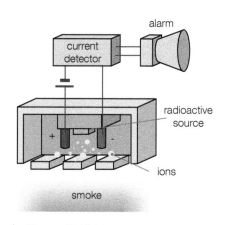

▲ **Figure 7.20** How a domestic smoke detector works

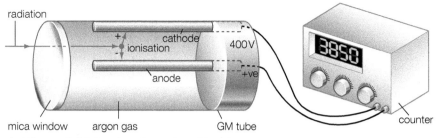

▲ **Figure 7.21** Section through a GM tube

It is not necessary for you to know how a GM tube works, but it is important to know how it could be used to do practical work on radiation.

Measuring the background radiation

First remove known sources of radiation from the laboratory, then set the GM counter to zero. Switch on the counter and start a stopwatch. After 30 minutes, read the count on the counter. Divide the count by 30 to obtain the background count rate in counts per minute. A typical figure is around 15 counts per minute. Fortunately, the background count in Northern Ireland does not present a serious health risk.

The background count must always be subtracted from any other count when measuring the activity from a specific source.

Safety precautions when using closed radioactive sources in schools

▶ Always store the sources in a lead-lined box, under lock and key, when not required for experimental use.
▶ Always handle sources using tongs, holding the source at arm's length and pointing it away from any bystander.
▶ Wear protective clothing.
▶ Always work quickly and methodically with sources to minimise the time of exposure and hence the dose to the user.

Measuring the approximate range of radiation

Alpha

▶ Place a GM tube on a wooden cradle and connect it to a counter.
▶ Hold an alpha source directly in front of the window of the tube and slowly increase the distance between the source and the tube. At about 3 cm (depending on the source used), the counter reading should fall dramatically to that of background radiation.
▶ Place a thin piece of paper in contact with the window of the GM tube. Bring the alpha source up to the paper so that the casing of the source touches it. The reading on the counter should now be the same as the background count, showing that the alpha particles are unable to penetrate the paper.

Beta

▶ Place a 1 mm thick piece of aluminium in contact with the window of the GM tube.

▶ Bring the beta source up to the aluminium so that the casing of the source touches it.

▶ The reading on the counter should be significantly above the background count, showing that some beta particles have penetrated the aluminium.

▶ Repeat the process with different sheets of aluminium, increasing the width by a millimetre at a time. At about 5 mm, there should be a significant reduction in the count rate, indicating the approximate range of beta particles in aluminium.

Gamma

If the beta particle experiment is repeated with a gamma source, there is practically no reduction in the count rate for a 5 mm thick piece of aluminium. If the aluminium sheets are replaced with lead, it will be found that even school sources will give gamma radiation that can easily penetrate several centimetres of lead (Figure 7.22).

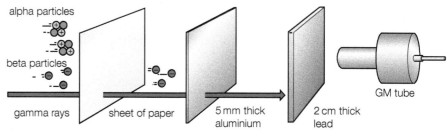

▲ **Figure 7.22** The penetrative range of the three types of radiation

Nuclear fission and nuclear fusion

Nuclear fission

Radioactivity involves the random disintegration of an unstable nucleus. Some heavy nuclei, like those of uranium, can actually be forced to split into two lighter nuclei. The process is called nuclear fission. This usually comes about as a result of the heavy nucleus being struck by a slow neutron (Figure 7.23a). The heavy nucleus splits and the fragments move apart at very high speed, carrying with them a vast amount of energy. At the same time, two or three fast neutrons are also emitted. These neutrons go on to initiate further fission, and so create a chain reaction.

Just how much energy is emitted in fission? The fission of a single uranium nucleus produces about 49 000 000 times more energy than would be produced by a single carbon atom (in coal) reacting with oxygen to produce carbon dioxide. It did not take long for physicists to realise the huge potential of energy production using nuclear fission.

In a nuclear power station, steps are taken to ensure that, on average, just one of the fission neutrons goes on to produce further fission. This is controlled nuclear fission. The heat produced in

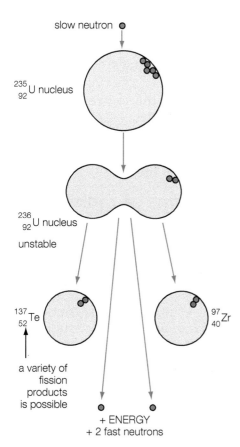

▲ **Figure 7.23a** The fission of a uranium-235 nucleus

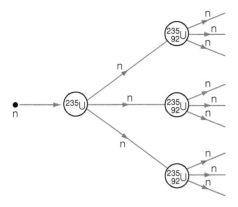

Figure 7.23b A chain reaction in uranium-235

the reaction is used to turn water into steam and drive a turbine to generate electricity. In a nuclear bomb, there is no attempt to control the fission process (Figure 7.23b).

A major disadvantage of all fission processes is that the fission fragments are almost always highly radioactive. This type of radioactive waste is extremely dangerous, and expensive measures must be taken to store it until the level of activity is sufficiently small. In some cases, this means the waste must be stored deep underground, in a vitrified (glass-like) state, for tens of thousands of years. The danger is that, over time, the containers may leak and cause underground water pollution.

A further danger comes from earthquakes. Earthquakes may rupture containers of radioactive waste buried underground, causing the radioactive material to leak into the soil and water systems. Even in Britain there are over 200 earthquakes every year, many so weak that they are barely recorded. But as recently as February 2008, there was an earthquake in Lincolnshire of magnitude 5.2. This earthquake lasted for roughly 10 seconds and caused some structural damage. The tremors were felt across a wide area of England and Wales, and as far west as Bangor, Northern Ireland.

Political, social, environmental and ethical issues relating to the use of nuclear energy to generate electricity

Nuclear energy is a controversial topic, which gives rise to strong arguments on both sides of the debate.

Arguments in favour of nuclear energy

▶ It can produce vast amounts of energy/electricity.
▶ It produces very little carbon dioxide (CO_2) and hence does not contribute to global warming.
▶ Nuclear energy provides the 'base-load' for national electricity generation, meaning that unlike some renewable energy sources, such as wind power, it is available at all times.
▶ It is a high density source of energy. 1 kg of uranium produces as much electricity as 20 000 kg of coal.
▶ It provides employment opportunities for many people.

The UK government and French energy giant EDF have signed a contract for the new £18bn Hinkley Point C nuclear power station. This will be the first new nuclear power station to be built in the UK since 1995.

Arguments against nuclear energy

▶ Disposing of nuclear waste can be dangerous and expensive.
▶ Many people are concerned about living close to nuclear power plants and the storage facilities used for radioactive waste.
▶ Past accidents have made people scared of nuclear power, including the disaster at Chernobyl in the Ukraine and the earthquake and tsunami in Fukushima, Japan in 2011, when several reactors were damaged leading to a meltdown and release of radiation.

In both cases, huge economic, health and environmental damage was caused to the area surrounding the power plants.

Although nuclear fission does not release carbon dioxide, the mining, transport and purification of the uranium ore releases significant amounts of greenhouse gases into the atmosphere.

Germany decided in 2011 to phase out all of its nuclear reactors by 2022, prepared instead to invest heavily in renewable energy.

Nuclear fusion

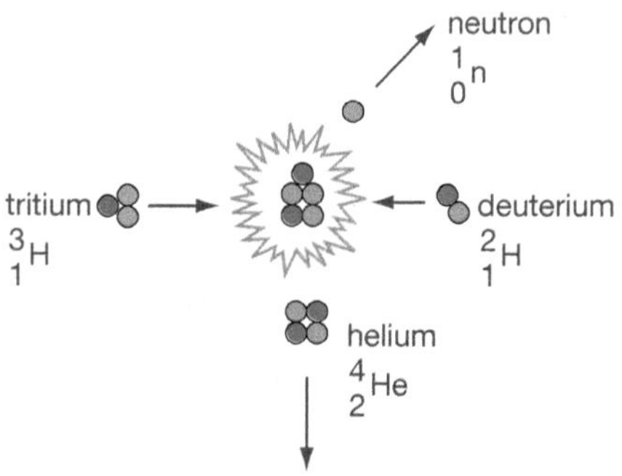

This is the process that goes on in stars like our Sun. At the centre of the Sun, the temperature is about 15 000 000 °C. At this temperature, the nuclei of atoms in a gas are all stripped of their orbiting electrons, and they move at a very high speed. These charged gas nuclei form a state of matter called a plasma. Being positively charged, the nuclei would normally repel each other, but if they are moving fast enough, they can join (or fuse) to form a new nucleus.

In the Sun, hydrogen isotopes known as deuterium (hydrogen-2) and tritium (or hydrogen-3) collide and fuse to create a new nucleus, helium-4 (Figure 7.24). This causes the release of a vast amount of energy, some of which eventually reaches Earth as electromagnetic radiation.

The equation representing this process is:

$$^2_1\text{H} + {}^3_1\text{H} \rightarrow {}^4_2\text{He} + {}^1_0\text{n} + \text{energy}$$

There have been many attempts to obtain controlled nuclear fusion on Earth. The world record for fusion power is held by the European reactor JET. In 1997, JET produced 16 MW of fusion power from a total input power of 24 MW.

Difficulties with fusion

The main problem is how to contain the reacting plasma at a high enough temperature and for a sufficiently long time for the reaction to take place. If we learn how to control nuclear fusion here on Earth, the rewards will be enormous. A large advantage of fusion is that the isotopes of hydrogen, deuterium and tritium (the more favoured isotopes of hydrogen) are widely available as the constituents of sea water, and so are nearly inexhaustible. Furthermore, fusion does not emit carbon dioxide or other greenhouse gases into the atmosphere, since its major by-product is helium, an inert non-toxic gas.

Fusion versus fission

Fusing nuclei together in a controlled way releases four million times more energy per kilogram than a chemical reaction such as the burning of coal, oil or gas, and fusing nuclei together in a controlled way releases four times as much energy as nuclear fission.

Is fusion a solution to the world's energy crisis?

There are many difficulties to overcome before nuclear fusion could provide electricity on a commercial scale, and it may be another 50

▲ **Figure 7.24** The fusion of deuterium and tritium

years before that happens. Nuclear fusion reactors will be expensive to build, and the system used to contain the plasma will be equally expensive because of the very high temperatures needed for the nuclei to fuse.

What is ITER?

ITER stands for International Thermonuclear Experimental Reactor.

In southern France, 35 nations are collaborating to build the world's largest tokamak. A tokamak is a toroidal (ring doughnut-shaped) container kept within a vacuum. The plasma needed for fusion to occur is stored within the tokamak using powerful magnetic fields.

For fusion to occur, a very large amount of energy needs to be concentrated in a very small space. ITER is intended to be the first fusion device to produce net energy, meaning that more energy should be produced from the fusion reaction than is needed to start the reaction. ITER is also designed to maintain fusion for long periods of time, and to test the integrated technologies and materials necessary for the commercial production of fusion-based electricity.

Thousands of engineers and scientists have contributed to the design of ITER since the idea for an international joint experiment in fusion was first launched in 1985. The ITER Members – China, the European Union, India, Japan, Korea, Russia and the United States – are now engaged in a 35-year collaboration to build and operate the ITER experimental device.

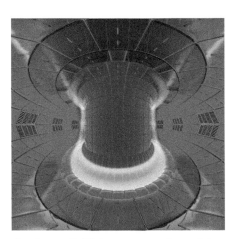

▲ **Figure 7.25** A cut-away view of the proposed ITER tokamak, revealing the doughnut-shaped plasma container inside the vacuum vessel

What will ITER do?

The amount of fusion energy a tokamak is capable of producing is a direct result of the number of fusion reactions taking place in its core. Scientists know that the larger the vessel, the larger the volume of the plasma, and therefore the greater the potential for fusion energy.

With ten times the plasma volume of the largest machine operating in 2016, the ITER tokamak could produce up to 500 MW of fusion power.

In 1997, JET produced 16 MW of fusion power from a total input power of 24 MW (see page 92). ITER is designed to produce 500 MW of fusion power from 50 MW of input power. ITER will not be able to capture the energy it produces as electricity, but it should prepare the way for a machine that can.

This section on ITER is not in the Double Award specification.

1 a) The table below shows the particles that make up a neutral carbon atom. Copy and complete the table showing the mass, charge, number and location of the particles. Some information has already been added to the table.

Particle	Mass	Charge	Number	Location
electron		−1		
neutron	1		6	in the nucleus
proton			6	

(7 marks)

b) Radon is a naturally occurring radioactive gas.
  i) Explain what is meant by radioactive. *(2 marks)*
  ii) Explain the danger of breathing radon gas into the lungs. *(2 marks)*
  iii) Explain the meaning of isotope in terms of the particles that make up the nucleus. *(2 marks)*

c) A student investigates the decay of a radioactive substance. She measures the corrected count rate of the substance every 20 minutes. The half-life of the substance is 20 minutes. At the start, the count rate was 800 counts per minute.
  i) Copy the graph grid shown in Figure 7.26 and plot this point and four more points that she found. *(5 marks)*

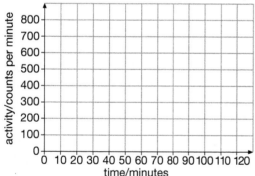

Figure 7.26

  ii) Draw a smooth curve through the plotted points. *(1 mark)*

d) i) The range of beta radiation in aluminium is several millimetres. Explain what this means. *(1 mark)*
  ii) Draw a neat and well-labelled diagram of the assembled apparatus that could be used to measure the range of beta particles in aluminium. *(3 marks)*
  iii) What measurements would be taken during this experiment? *(3 marks)*
  iv) How would you use these measurements to find the range of beta particles in aluminium? *(2 marks)*
  v) Sketch the graph that you would expect to obtain from these measurements and mark on it the range of the beta radiation. Label each axis. *(2 marks)*

2 $^{12}_{6}\text{C}$ and $^{14}_{6}\text{C}$ are both isotopes of carbon.
  a) i) Write down one similarity of the nuclei of these isotopes. *(1 mark)*
  ii) Write down one difference between the nuclei of these isotopes. *(1 mark)*
  b) $^{14}_{6}\text{C}$ is radioactive. It decays to nitrogen by emitting a beta particle. Complete the equation below to describe the decay.
  $$^{14}_{6}\text{C} \rightarrow {}^{A}_{Z}\text{N} + {}^{0}_{-1}\text{e}$$ *(2 marks)*
  c) $^{14}_{6}\text{C}$ is present in all living materials and in all materials that have been alive. It decays with a half-life of 6000 years.
  i) Explain the meaning of the term 'a half-life of 6000 years'. *(2 marks)*
  ii) The activity of a sample of wood from a freshly cut tree is measured to be 80 disintegrations per second. Estimate the decrease in activity of the sample after 3 half-lives. *(2 marks)*

3 A radioactive isotope of gold emits gamma rays. It is injected into a patient's bloodstream and used to study the working of the patient's heart. The gamma radiation emitted by the gold is detected outside the patient's body by a device called a gamma camera.

a) Why would a radioactive isotope that emits alpha radiation be unsuitable for this purpose? *(2 marks)*

To check the half-life of this isotope of gold, a radiographer measured the activity of a sample of the isotope every 10 s. He then corrected for the background activity. His measurements are shown in the table below.

Corrected activity/ counts per second	400	320	250	198	160	100	80
Time/seconds	0	10	20	30	40	60	70

b) What causes background activity and how did the radiographer correct his measurements? *(2 marks)*

c) Using the measurements above, plot a graph of corrected activity (y-axis) against time (x-axis). *(5 marks)*

d) Use the graph to find the half-life of this isotope of gold. Show clearly how you use the graph to obtain the best value of this half-life. *(2 marks)*

4 A radioactive decay series can be represented on a graph of mass number, A, against atomic number, Z. Part of a table for such a series is given below:

Element (symbol)	Atomic number	Mass number	Decays by emitting	Leaving element
U	92	238	α	Th
Th	90	234	β	Pa
Pa	91	234	β	
	92	234	α	
	90	230		Ra
Ra	88	226		Rn
Rn	86			Po
Po		218	α	Pb
Pb				Bi
Bi	83			Po

a) In what ways do mass number and atomic number change in
i) α decay
ii) β decay? *(4 marks)*

b) Copy and complete the table. *(7 marks)*

c) Identify two pairs of isotopes using the table. *(2 marks)*

5 A sample containing 100 grams of a uranium isotope arrives at a factory. The table below shows how the mass of the isotope changes over time.

Mass of isotope in grams	Time in days
100	0
72	10
52	20
37	30
27	40

a) Explain the meaning of the following words:
i) half-life
ii) isotope. *(2 marks)*

b) Plot a graph of mass of isotope (y-axis) against time (x-axis). *(4 marks)*

c) From the graph, calculate as accurately as possible the half-life of this isotope. *(2 marks)*

d) Estimate the mass of the uranium present in the sample three weeks before it arrived in the factory. *(2 marks)*

6 A certain material has a half-life of 12 minutes. What proportion of that material would you expect still to be present an hour later? *(3 marks)*

7 A detector of radiation is placed close to a radioactive source that has a very long half-life. In four consecutive 10 second intervals, the following number of counts were recorded: 100, 107, 99, and 102. Why were the four counts different? *(2 marks)*

8 a) The full symbol for a nucleus of carbon-14 is $^{14}_{6}C$. Copy and complete the table below by naming the particles in a nucleus of carbon-14 and give the quantity of each in the nucleus of carbon-14.

Particle	Quantity in the nucleus

(4 marks)

b) Four unknown nuclei are labelled X1, X2, X3 and X4. Their full symbols are given below.

$^{30}_{15}X1$ $^{30}_{16}X2$ $^{32}_{17}X3$ $^{33}_{16}X4$

i) Which, if any, of these nuclei are isotopes of the same element? *(1 mark)*

ii) Explain your answer. *(1 mark)*

9 Fission and fusion are nuclear reactions. The table below is intended to show a number of significant differences between the two reactions. Copy and complete the table using the following list.
1. building of larger nuclei from small nuclei
2. splitting up of large nuclei
3. nuclear power station
4. requires very high temperatures to start
5. the Sun
6. hydrogen
7. uranium
8. will start at normal temperatures

Nuclear reaction	Fusion	Fission
Where the process can be found happening		
Fuel used		
Description of the reaction		
Conditions required to starts		

(4 marks)

10 Cobalt-60 is a beta emitter which decays to nickel. The nickel produced is a gamma emitter.
a) Copy and complete the decay equation below for the full beta decay process by writing the correct values in the boxes.

$$^{60}_{27}\text{Co} \rightarrow \boxed{}\text{Ni} + ^{0}_{-1}\text{e}$$ (2 marks)

After 15 years, the measured activity of a cobalt-60 source is found to have fallen from 120 counts per minute to 15 counts per minute.
b) What is the half-life of cobalt-60? (3 marks)
c) State two possible uses of gamma radiation. (2 marks)

11 Uranium is an alpha emitter.
a) Explain why wearing heavy gloves to handle uranium rods could be sufficient protection. (1 mark)
b) State two alternatives to specialised clothing as ways of protecting people from the radiation from radioactive sources. (2 marks)

12 a) Describe the process of nuclear fusion. Your description should include:
- the particles involved
- what happens when nuclear fusion takes place
- where nuclear fusion occurs naturally. (6 marks)
b) A great deal of money is being invested on research into nuclear fusion.
i) Suggest a reason why. (1 mark)
ii) Give two practical difficulties which must be overcome before fusion reactors become viable. (2 marks)

c) i) Explain the need for the ITER project. (3 marks)
ii) How is ITER being implemented? (2 marks)

13 Figure 7.27 shows the decay curve for a sample of iodine-123.

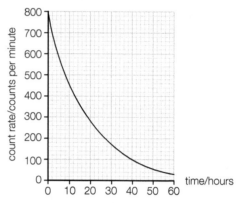

Figure 7.27 Decay curve for iodine-123

a) i) Using the graph, write down the half-life of iodine-123. (1 mark)
ii) What is the count-rate after 24 hours? (1 mark)
b) i) If 0.08 g of iodine-123 in solution is injected into a patient, what is the maximum amount still in the body after 39 hours? (1 mark)
ii) Why is it unlikely to actually be this much? (1 mark)
c) Iodine-123 is a gamma emitter. Another isotope, iodine-131, which has a half-life of 8 days, is a beta emitter. Give two reasons why iodine 131 is not suitable as a tracer in the body. (2 marks)

14 Radioactive sources can be used to treat a person with a cancerous tumour inside their body. The radioactive source is held directly over the tumour, as shown in Figure 7.28.

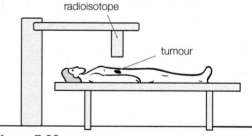

Figure 7.28

a) What type of radiation should the source emit if it is to penetrate the body and reach the tumour? (1 mark)
b) Why is this type of radiation effective in the treatment of tumours? (1 mark)
c) What disadvantages are there to using radioactivity to treat the tumour? (1 mark)
d) What is meant by the half-life of a radioactive source? (2 marks)
e) Should the radioactive source used in the treatment of the tumour have a short or long half-life? Explain your answer. (2 marks)

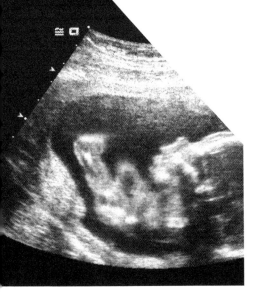

Types of waves

Waves transfer energy from one point to another, but they do not, in general, transfer matter. Radio waves, for example, carry energy from a radio transmitter to your home, but no matter moves permanently in the air as a result.

All waves are produced as a result of vibrations, and they can be classified as longitudinal or transverse. A vibration is a repeated movement, first in one direction and then in the opposite direction.

Longitudinal waves

A longitudinal wave is one in which the particles vibrate **parallel** to the direction in which the wave is travelling. The only types of longitudinal waves relevant to your GCSE course are:

▶ sound waves
▶ ultrasound waves
▶ slinky spring waves
▶ P-type earthquake waves.

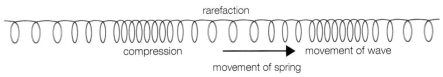

rarefaction

compression → movement of wave

movement of spring

⌃ **Figure 8.1** A longitudinal wave moving along a slinky spring

It is easy to demonstrate longitudinal waves by holding a slinky spring at one end and moving your hand backwards and forwards parallel to the axis of the stretched spring. Compressions are places where the coils (or particles) bunch together. Rarefactions are places where the coils (or particles) are furthest apart.

All longitudinal waves are made up of compressions and rarefactions. In the case of sound waves, the particles are the molecules of the material through which the sound is travelling. These molecules bunch together and separate just as they do in a longitudinal wave on a slinky spring.

Transverse waves

A transverse wave is one in which the vibrations are **at 90°** to the direction in which the wave is travelling. Most waves in nature are transverse – some examples are:

▶ water waves (Figure 8.2)
▶ slinky spring waves (Figure 8.3)
▶ waves on strings and ropes
▶ electromagnetic waves.

A transverse wave pulse can be created by shaking one end of a slinky. The pulse moves along the slinky, but the final position of the slinky is exactly the same as it was at the beginning (Figure 8.3). None of the material of the slinky has moved permanently. But the wave pulse has carried energy from one point to another.

Figure 8.2 represents water waves. You can see that water waves are transverse. A cork floating on the surface of some water bobs up and down as the waves pass. The vertical vibration of the cork is perpendicular to the horizontal motion of the wave. Energy is transferred in the direction in which the wave is travelling.

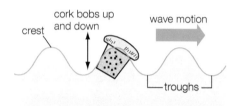

▲ **Figure 8.2** Transverse water waves move the cork up and down

> **Tip**
>
> Slinky springs can be used to demonstrate both longitudinal and transverse waves, so it is best not to use these as an example of a longitudinal (or a transverse) wave.

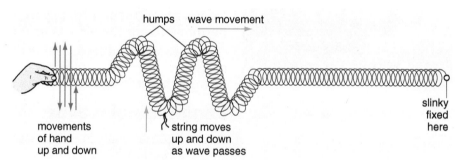

▲ **Figure 8.3** Transverse waves travelling through a slinky

There are many other examples that show that waves carry energy:

▶ Visible light, infrared radiation and microwaves all make things heat up.
▶ X-rays and gamma waves can damage cells by disrupting DNA.
▶ Loud sound waves can cause objects to vibrate (for example, your eardrum).
▶ Water waves can be used to generate electricity.

Describing waves

There are a number of important definitions relating to waves that must be learned.

The frequency of a wave is the number of complete waves passing a fixed point in a second. Frequency is given the symbol f, and is measured in units called hertz (abbreviation Hz).

Sometimes the units kHz and MHz are used. You should remember that $1\,\text{kHz} = 1000\,\text{Hz}$ and $1\,\text{MHz} = 1000\,\text{kHz} = 1\,000\,000\,\text{Hz}$.

The wavelength of a wave is the distance between two consecutive crests or troughs. Wavelength is given the symbol λ, and is measured in metres. λ is the Greek letter 'l' and is pronounced 'lamda'.

The amplitude of a wave is the greatest displacement of the wave from its undisturbed position. Amplitude is measured in metres. The wavelength and amplitude for a transverse wave are illustrated in Figure 8.4.

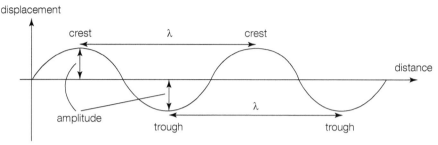

▲ **Figure 8.4** Displacement–distance graph to illustrate wavelength and amplitude

Wavelength and amplitude of longitudinal waves

It is much easier to visualise wavelength and amplitude for transverse waves than for longitudinal waves. For a longitudinal wave, the wavelength is the distance between the centre of one compression and the next (Figure 8.5).

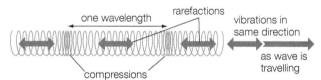

▲ **Figure 8.5** In longitudinal waves, the vibrations are along the same direction as the wave is travelling

But what is the amplitude of a longitudinal wave? Remember that the particles in a longitudinal wave vibrate backwards and forwards parallel to the direction in which the wave is moving. The amplitude of a longitudinal wave is the maximum distance a particle moves from the centre of this motion (Figure 8.6).

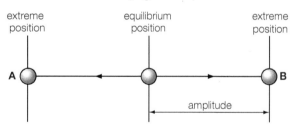

▲ **Figure 8.6** The amplitude of a longitudinal wave

The wave equation

Imagine a wave with wavelength λ (metres) and frequency λ (hertz). From the definition of frequency, f waves pass a fixed point in 1 second. Each wave has a length λ, so, the total distance travelled every second is $f \times \lambda$.

The distance travelled in a second is the speed. So:

$$\text{wave speed} = \text{frequency} \times \text{wavelength}$$

$$v = f \times \lambda$$

Table 8.1

Frequency	Wavelength	Speed
Hz	cm	cm/s
	m	m/s
	km	km/s

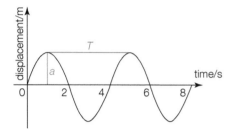

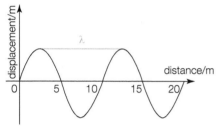

▲ **Figure 8.7** Graphs showing displacement against time and displacement against distance

This is an important equation that you should memorise and learn how to use.

Note that the units used in the wave equation must be consistent, as shown in Table 8.1.

Graphs and waves

We can represent waves on graphs like those shown in Figure 8.7.

Notice that the upper graph is displacement against time. The lower graph is displacement against distance. The vertical axis on both graphs is displacement, so we can find the amplitude from either graph.

The red line in the upper graph shows the time, T, between the crests passing a fixed point. This time is known as the period. Students often wrongly think T is a distance (the wavelength).

The blue line in the lower graph shows the distance between consecutive crests. This is the wavelength, λ.

The period, T, in Figure 8.7 is 4 seconds, and the wavelength, λ, is 10 metres.

How could we find the speed from these data?

The graphs tell us that the wave is travelling 10 metres in 4 seconds. We know that speed = $\dfrac{\text{distance}}{\text{time}}$.

So, the wave speed = $\dfrac{10\,\text{m}}{4\,\text{s}}$

$= 2.5\,\text{m/s}$

We can also find the frequency from the top graph. The period (the time taken for one wave to pass), T, is 4 seconds, so 0.25 waves must pass every second.

So the frequency, $f = \dfrac{1}{T}$

$= 0.25\,\text{Hz}$

We can use the wave equation to confirm the speed:

$v = f \times \lambda$

$v = 0.25\,\text{Hz} \times 10\,\text{m}$

$= 2.5\,\text{m/s}$

> **Test yourself** ✎
>
> 4 Copy and complete the table below. Note carefully the unit in which you are to give your answers. The first one has been done for you.
>
> | Wavelength | Frequency | Speed |
> |------------|-----------|-------|
> | 5 m | 200 Hz | 1000 m/s |
> | 12 m | 50 Hz | .... m/s |
> | 3 m | 60 Hz | .... m/s |
> | ...... m | 4 Hz | 20 cm/s |
> | ...... m | 5 Hz | 2.5 km/s |
> | 16 m | ...... Hz | 80 cm/s |
> | 6×10^4 m | ...... Hz | 3×10^8 m/s |

5 The vertical distance between a crest and a trough is 24 cm, and the horizontal distance between the first and the fifth wave crests is 40 cm. If 30 such waves pass a fixed point every minute, find the amplitude, frequency, wavelength and speed of the waves.

Plane wavefronts

We can learn much about the behaviour of waves using a ripple tank, as shown in Figure 8.8.

A motor makes a straight dipper vibrate up and down continuously. This produces straight water waves, also called plane waves. By shining a light from above the tank, we can see bright and dark patches on the screen below. These patches show the wave crests and troughs. The direction of movement of the water waves is always at right angles to the wave fronts.

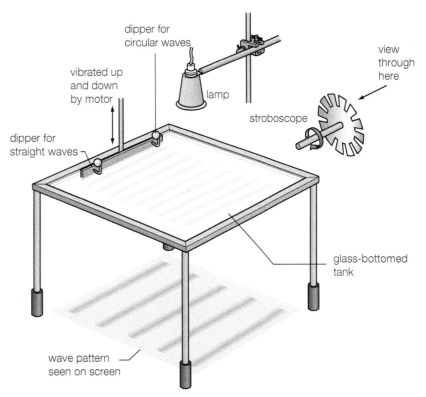

▲ **Figure 8.8** Using a ripple tank to observe waves

Water waves move quite quickly, and it can sometimes be hard to see what is happening.

Looking through a rotating stroboscope can make the waves appear to stand still.

Suppose the stroboscope had 12 slits and rotated twice every second. The tank could be seen 24 times a second. If the waves had a frequency of 24 Hz, every $\frac{1}{24}$th of a second each wave would have moved forward by exactly one wavelength. This would mean that when viewed through the stroboscope, the wave pattern would appear to be stationary.

Reflection

Figure 8.9 shows some plane waves approaching a straight metal barrier. The barrier is big enough to prevent waves going 'over the top'. The incident waves are reflected from the barrier.

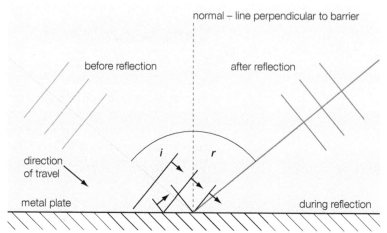

▲ **Figure 8.9** The reflection of waves off a plane surface. The angle of incidence = i, the angle of reflection = r

Note that:

▶ the angle of incidence always equals the angle of reflection
▶ the wavelengths of incident and reflected waves are equal
▶ the frequency of the incident waves is the same as that of the reflected waves
▶ there is continuity of incident waves and reflected waves at the barrier.

The behaviour of water waves at a boundary is very similar to that of light at a mirror. However, water waves can be observed easily because they have a wavelength of many centimetres. Light waves have a wavelength typically around half a millionth of a metre, so their wave behaviour is more difficult to demonstrate.

Refraction

In Figure 8.10, water waves are travelling from deep water to shallow water. A region of shallow water in a ripple tank can be made by immersing a rectangular glass block. The block displaces the water so that the water directly above it is shallow while the surrounding water is deeper.

Waves travel more slowly in shallow water than they do in deep water. This change of speed is called refraction. Since the same number of waves leave the deep water as enter the shallow water every second, the frequencies in the deep and shallow regions must be the same. This means that the waves in shallow water must have a shorter wavelength than those in deep water.

When water waves enter the shallow region obliquely (at an angle), they not only slow down, but also change direction, as shown in Figure 8.11.

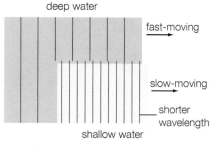

▲ **Figure 8.10** Water waves travelling from deep water to shallow water

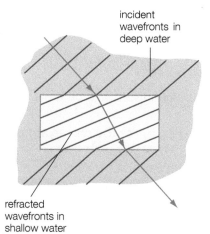

▲ **Figure 8.11** Waves changing direction as they pass from deep to shallow water

Note that:

▶ The angle of incidence in deep water is always bigger than the angle of refraction in shallow water.

▶ The wavelength and speed of waves in deep water are greater than those waves in shallow water.

▶ The frequencies of waves in both deep and shallow water are the same.

▶ There is continuity of incident and refracted waves at the boundary.

Analogy between the behaviour of water waves and the behaviour of light

There are many similarities between the behaviour of water waves in a ripple tank and the behaviour of light. Both show reflection, in which the angle of incidence is equal to the angle of reflection. When water waves travel from deep water into shallow water, they refract in exactly the same way as when light travels from air into glass. These similarities mean that we can say there is an analogy between light waves and water waves.

However, we can see much more happening when we look at the reflection and refraction of water waves than we can with light. For example, we can see that the wavelengths, frequencies and speed of the incident water waves are equal to the wavelengths, frequencies and speed of the reflected water waves. This led physicists to suspect that the wavelength, frequency and speed of light waves do not change when they reflect, and this later proved to be the case.

By looking at the movement of floating chalk dust on the surface of water, we can say with confidence that water waves are transverse. This leads us to think that light waves are transverse too – and, as we know, this is the case.

But what does refraction tell us? When water waves refract as they pass from deep water to shallow water, they bend towards the normal. This is exactly what happens when light passes from air into glass. When we measure the speed of water waves, we find that deep water waves move faster than shallow water waves. If we apply the analogy again, then we would suspect that light travelling in air is moving faster than light travelling in glass. Once again, this turns out to be the case. The analogy is summarised in Table 8.2.

Example

A deep water wave of wavelength 12 cm and speed 36 cm/s enters a shallow region where the wavelength is 8 cm. Find the frequency and wave speed in the shallow water.

Answer

frequency in shallow water = frequency in deep water

$$= \frac{v}{\lambda}$$

$$= \frac{36 \, cm/s}{12 \, cm}$$

$$= 3 \, Hz$$

speed in shallow water $= f \times \lambda$

$$= 3 \, Hz \times 8 \, cm$$

$$= 24 \, cm/s$$

Table 8.2 Analogy between water waves and light

Water waves	Light
When they reflect:	When it reflects:
angle of incidence = angle of reflection	angle of incidence = angle of reflection
reflected wavelength = incident wavelength	reflected wavelength = incident wavelength
reflected frequency = incident frequency	reflected frequency = incident frequency
reflected speed = incident speed	reflected speed = incident speed
When they pass from deep into shallow water, they bend towards the normal.	When it passes from air into glass and it refracts, it bends towards the normal.
The refracted wavelength is less than the incident wavelength. refracted frequency = incident frequency	The refracted wavelength is less than the incident wavelength. refracted frequency = incident frequency
The refracted speed is less than the incident speed.	The refracted speed is less than the incident speed.

Test yourself

6 Figure 8.12 shows three wavefronts in a ripple tank approaching a solid barrier. The barrier acts as a reflector.
   a) Copy and complete the diagram to show what happens when the waves are reflected.
   b) In what way, if at all, do the frequency, wavelength and speed of water waves change when they are reflected?

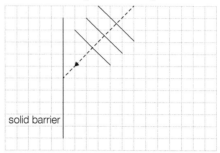

▲ **Figure 8.12** Three wavefronts in a ripple tank

7 Figure 8.13 shows three wavefronts in deep water in a ripple tank. The water to the left of the vertical line is shallow water.
   a) Copy and complete the diagram to show what happens when the waves are refracted.
   b) In what way, if at all, do the frequency, wavelength and speed of these water waves change when they are refracted?

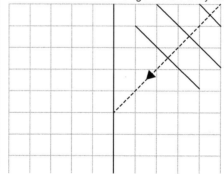

▲ **Figure 8.13** Three wavefronts in deep water in a ripple tank

8 a) The behaviour of water waves is analogous to that of light. What does this mean?
   b) State three properties of water waves that are analogous to those of light waves.

Echoes in sound and ultrasound

Like all waves, sound and ultrasound can be made to reflect (Figure 8.14).

This happens in a similar way to the reflection of water waves. As with water waves, the angle of incidence is always equal to the angle of reflection.

Audible sound ranges in frequency from 20 Hz to 20 000 Hz. Sound above 20 000 Hz is called ultrasound and cannot be heard by humans. It can, however, be detected by bats, dogs, dolphins and many other animals.

▲ **Figure 8.14** The reflection of sound against a hard surface

Reflected sound (and ultrasound) is called an echo. Humans have found clever ways to use ultrasound echoes:

▶ scanning metal castings for faults or cracks (for example in rail tracks)

▶ scanning a pregnant woman's womb to check on the development of her baby

▶ scanning soft tissues to diagnose cancers

▶ locating fish by seagoing trawlers

▶ mapping the surface of the ocean floor in oceanography.

We will look at some of these applications in a little more detail.

An application of ultrasound in medicine

In an ultrasound scan of an unborn baby, a probe is moved across the mother's abdomen. The probe sends out ultrasound waves and also detects the reflections. The low wavelength of the ultrasound waves means that, unlike audible sound waves, ultrasound can be sent out in a very narrow beam and focused on the unborn baby. The other end of the probe is connected to a computer.

By examining the reflected waves from the womb, the computer builds up a picture of the foetus (unborn baby) like that in Figure 8.15. Unlike X-rays, ultrasound is now known to be quite safe for this purpose.

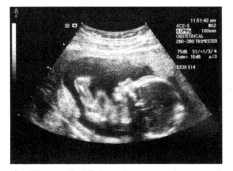

▲ **Figure 8.15** An ultrasound scan of an unborn baby

Ultrasound can also be used to measure the diameter of the head of the baby as it develops in the womb (Figure 8.16). When the ultrasound reaches the baby's head at A, some ultrasound is reflected back to the detector and produces pulse A on the cathode ray oscilloscope (CRO). Some ultrasound passes through the head to point B, and is then reflected back to the detector. This reflection produces pulse B on the CRO.

In the diagram of the CRO screen, each horizontal division corresponds to a time of 40 microseconds (40 μs = 40 × 10⁻⁶ s).

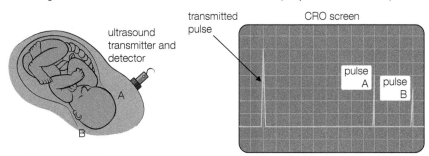

▲ **Figure 8.16** The diameter of a baby's head can be measured using ultrasound

The time interval between the arrival of pulse A and the arrival of pulse B at the detector corresponds to 3 divisions on the CRO. Since each division is 40 μs, this represents a total time of 120 μs. This additional 120 μs is the time taken for ultrasound to travel from A to B and back to A. The time for ultrasound to travel from B to A is therefore half of that or 60 μs.

Now, physicists know that ultrasound travels at a speed of 1500 m/s in a baby's head. So the width of the head can be found as follows:

width of head = speed × time

$$= 1500 \, \text{m/s} \times 60 \times 10^{-6} \, \text{s}$$
$$= 0.09 \, \text{m}$$
$$= 9 \, \text{cm}$$

Scanning metal castings

Railway tracks do not last forever. They wear out. It is important that we find out early if they are developing cracks or flaws below the surface.

Ultrasound scanners attached to specially fitted rail carriages (Figure 8.17) can be used to detect these cracks and flaws. Ultrasound passes through the track. If there is a crack or other flaw it can be imaged using the same science that allows ultrasound technicians to obtain a picture of a baby in the womb.

Alternatively, the depth to the crack can be found using the same science used to measure the diameter of the head of a baby in the womb. To carry out the measurement, we need to know that ultrasound travels at 5000 m/s in steel.

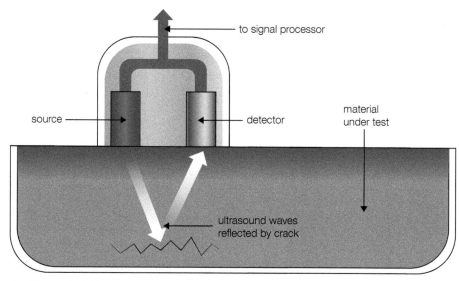

▲ **Figure 8.17** Detecting cracks in metals

Sonar

Sonar stands for **SO**und **Na**vigation **A**nd **R**anging and was originally developed to detect submarines in the early twentieth century. We will look at how fishermen use sonar to detect shoals of fish and to measure how far they are below the surface.

The fishing trawler sends out an ultrasound pulse. Ultrasound travels at 1500 m/s in seawater. The ultrasound hits the fish and is reflected as an echo. This echo is detected back on the trawler. By measuring the time between the transmission of the sound and the detection of the echo, we can calculate the depth of the fish below the surface of the sea.

Example

A fishing trawler produces an ultrasound pulse and 0.4 s after it detects an echo reflected from a shoal of fish. Assuming the speed of ultrasound in seawater is 1500 m/s:

i) Calculate the total distance travelled by the ultrasound.

ii) Calculate the distance from the trawler to the shoal of fish.

iii) Explain why a second echo may be detected after the first.

Answer

i) total distance = speed × time

= 1500 m/s × 0.4 s

= 600 m

ii) distance from the trawler to the fish = ½ × total distance

= ½ × 600 m

= 300 m

iii) The first echo is from the fish. The second echo is from the sea bed, which is further away from the trawler than the fish, so the ultrasound from the sea bed takes longer to reach the trawler than the echo from the fish.

Radar

Radar stands for **RA**dio **D**etection **A**nd **R**anging and was originally developed during World War II to detect enemy aircraft and find their distance from the radar station.

Radar waves are in the microwave section of the electromagnetic spectrum. Think of a radar beam as a powerful beam of microwaves. They have wavelengths ranging from a few millimetres to just over a metre. Because radar waves are incredibly fast (3×10^8 m/s), they are used to track very fast objects which may be a large distance away. So, for example, they are used by air traffic controllers to track passenger airliners, by the military to track missiles, and by the coastguard to detect ships.

Radar cannot be used under water. The water absorbs the radar (microwaves) within a metre or so. In addition, radar cannot be used to measure very small distances. This is because radar is so fast that the time taken would be too small to measure easily.

However, the physics of radar is very similar to that of sonar. If you are asked to solve a mathematical problem involving radar, it is likely that numbers will be given in index form as shown in the example.

Electromagnetic waves

Electromagnetic waves are members of a family with common properties called the electromagnetic spectrum. They:

▶ can travel in a vacuum (unique property of electromagnetic waves)
▶ all travel at exactly the same speed in a vacuum
▶ are transverse waves.

Electromagnetic waves also show properties common to all types of wave. They:

▶ carry energy
▶ can be reflected
▶ can be refracted.

We give names to seven broad sections of the electromagnetic spectrum, according to their wavelengths and the effects they produce. The properties of electromagnetic waves depend very much on their wavelength. In Table 8.3 they are arranged in order of increasing wavelength (or decreasing frequency). You need to be able to list these waves in order of increasing (and decreasing) wavelength, but do not need to remember specific wavelengths!

Table 8.3 The electromagnetic spectrum

Electromagnetic wave	Typical wavelength
Gamma (γ) rays	0.01 nm
X-rays	0.1 nm
Ultraviolet light	10 nm
Visible light	500 nm
Infrared light	0.01 mm
Microwaves	3 cm
Radio waves	1000 m

1 nanometre (nm) = 1×10^{-9} m

Example

The diagram shows a ground-based radar station A. It transmits a radar beam which reflects off aircraft B. The radar echo is received at A 2.2×10^{-4} s after the original radar transmission. If the speed of radar waves is 3×10^8 m/s, calculate the distance AB in kilometres.

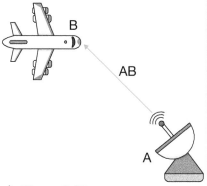

▲ **Figure 8.18**

Answer

The time taken for radar to travel from A to B and back to A is 2.2×10^{-4} s.

So, the time taken for radar to travel from A to B is
$\frac{1}{2} \times 2.2 \times 10^{-4}$ s $= 1.1 \times 10^{-4}$ s.

Distance AB = speed × time

$= 3 \times 10^8$ m/s $\times 1.1 \times 10^{-4}$ s

$= 3.3 \times 10^4$ m

$= 33$ km

Dangers of electromagnetic waves

Table 8.4

Electromagnetic wave	Dangers
Gamma (γ) rays	damage cells and disrupt DNA, which may lead to cancer
X-rays	damage cells and disrupt DNA, which may lead to cancer
Ultraviolet light	certain wavelengths can damage skin cells, disrupting DNA and potentially leading to skin cancer
Visible light	intense visible light can damage the eyes (for example in snow blindness)
Infrared light	felt as heat and can cause burns
Microwaves	cause internal heating of body tissues which, some say, can lead to eye cataracts
Radio waves	large doses of radio waves are believed to cause cancer, leukaemia and other disorders, and some people claim the very low frequency radio waves from overhead power cables near their homes have affected their health

Test yourself

9 Copy the diagram below and list the sections of the electromagnetic spectrum in order of increasing frequency by writing their names in the boxes. One box has been completed for you.

			visible light			

increasing frequency →

10 State a property unique to electromagnetic waves.

11 Below are three sections of the electromagnetic spectrum, not arranged in any particular order.

gamma rays, radio waves, ultraviolet light

Copy and complete the table below writing the name of the missing electromagnetic wave opposite its typical wavelength.

Hint: First identify the missing sections of the spectrum and write them in order of increasing wavelength.

Wave				
Typical wavelength/m	1×10^{-10}	6×10^{-7}	1×10^{-5}	1×10^{-3}

12 The emitter in Figure 8.19 sends out a pulse of sound. An echo from the object is detected after 2.5 ms. If sound travels at 340 m/s in the air, calculate the distance marked d.

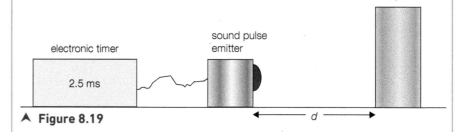

▲ **Figure 8.19**

1 State the difference between a transverse and a longitudinal wave and give an example of each type of wave. *(4 marks)*

2 What is meant by wavelength, frequency and amplitude when applied to longitudinal waves? *(3 marks)*

3 What physical property of a water wave never changes as a result of:
a) reflection and
b) refraction? *(2 marks)*

4 The vertical distance between a crest and a trough of a water wave in the mid-Atlantic is 16 m, and the horizontal distance between the first and the fifth wave crests is 80 m. If 12 such waves pass a fixed point every minute, find the amplitude, frequency, wavelength and speed of the waves. *(8 marks)*

5 A water wave has period $T = 0.25$ s, vertical distance from peak to trough = 25 cm and $\lambda = 1.2$ m. Find the amplitude, frequency and speed of the wave. *(6 marks)*

6 Figure 8.20 shows three wavefronts in shallow water in a ripple tank. The water to the left of the vertical line is deep water.
a) Copy and complete the diagram to show what happens when the waves are refracted. *(3 marks)*
b) In what way, if at all, do the frequency, wavelength and speed of these water waves change when they are refracted? *(3 marks)*

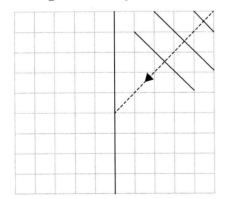

Figure 8.20

7 a) What is ultrasound? *(1 mark)*
b) In what way is ultrasound different from the sound of human speech? *(1 mark)*

8 State two differences between ultrasound waves and electromagnetic waves. *(2 marks)*

9 a) List the members of the electromagnetic spectrum in order of decreasing wavelength. *(2 marks)*
b) Which three members of the electromagnetic spectrum have been proven to cause cancer? *(3 marks)*

10 State the difference between a transverse and a longitudinal wave and give an example of each type of wave. *(4 marks)*

11 When the frequency of sound is changed, the wavelength also changes. The table below shows the results of an experiment to measure the wavelength of sound at different frequencies. The unit for $1/\lambda$ is 1/m.

Wavelength, λ/m	0.7	1.0	1.5	2.5	4.0
Frequency, f/Hz	460	320	210	130	80
$\frac{1}{\lambda}$/1/m				0.40	0.25

a) Complete the table by entering the missing numbers in the third row. *(3 marks)* Two entries have already been done for you.
b) Plot (on graph paper) the graph of f/Hz on the vertical axis against $\frac{1}{\lambda}$/1/m on the horizontal axis and draw the straight line of best fit. *(4 marks)*
c) Find the gradient of your line of best fit and state its unit. *(3 marks)*
d) What is the physical significance of the gradient of the line of best fit? *(1 mark)*
e) Use your graph to find the wavelength of sound of frequency 250 Hz. *(2 marks)*

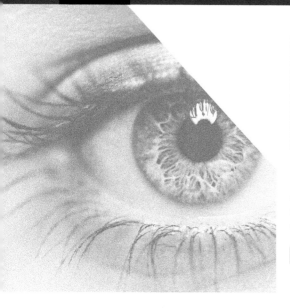

Reflection of light

All of us are familiar with the way light reflects from a straight (plane) mirror.

The essential ideas are shown in Figure 9.1.

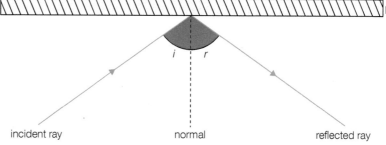

▲ **Figure 9.1** Reflection by a plane mirror

▶ The ray of light travelling towards the mirror is called the **incident ray**.

▶ The perpendicular to the mirror where the incident ray strikes it is called the **normal**.

▶ The ray that travels away from the mirror is called the reflected ray.

▶ The angle between the incident ray and the normal is called the angle of incidence, i.

▶ The angle between the reflected ray and the normal is called the angle of reflection, r.

Experiments show that the angle of incidence is always equal to the angle of reflection. This is known as the law of reflection. You should be able to describe the following experiment to demonstrate this law.

Demonstrating the law of reflection

We can use apparatus such as that shown in Figure 9.2 to demonstrate the law of reflection.

▶ With a sharp pencil and a ruler, draw a straight line AOB on a sheet of white paper using a ruler.

▶ Use a protractor to draw a normal, N, at point O (90° to the mirror).

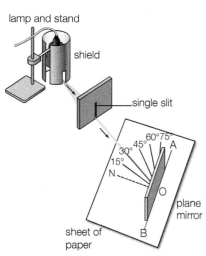

▲ **Figure 9.2** Single slit test illustrating the law of reflection

▶ With the protractor, draw straight lines at various angles to the normal ranging from 15° to 75°.

▶ Place a plane mirror on the paper so that its back rests on the line AOB as in Figure 9.2.

▶ Using a ray box, shine a ray of light along the line marked 15°.

▶ Mark two crosses on the reflected ray on the paper.

▶ Remove the mirror and, using a ruler, join the crosses on the paper with a pencil. Extend the line backwards to point O – this line shows the reflected ray.

▶ Measure the angle of reflection with a protractor.

▶ Record the angles of incidence and reflection in a table.

▶ Repeat the experiment for different angles of incidence up to 75°.

Note that when light is reflected off a rough surface, such as paper, the law of reflection still applies. However, because the paper surface is rough (Figure 9.3b), the light is reflected in many different directions and cannot produce a clear image. This is called diffuse reflection.

Where is the image we see in a plane mirror? We know it is behind the mirror, but how far behind?

To answer that question, we need to do another experiment.

Where is the image?

We can use apparatus such as that shown in Figure 9.3 to locate the image produced by a plane mirror.

▶ Support a plane mirror vertically on a sheet of white paper placed on a horizontal surface, and with a pencil, draw a straight line at the back to mark the position of the reflecting surface.

▶ Use a ray box to direct two rays of light from point O towards points A and B on the mirror as in Figure 9.4.

▶ Mark the position of point O with a cross using a pencil.

▶ Mark two crosses on each of the real reflected rays.

▶ Remove both the ray box and the mirror.

▶ Using a ruler, join the crosses with a pencil line to obtain the paths of the real rays from A and B.

▶ Extend these lines behind the mirror (these are called virtual rays) – they meet at I, the point where the image was formed.

▶ Measure the distance from the image I to the mirror line (IN), and the distance from the object O to the mirror line (ON). They should be the same.

▶ Repeat the experiment for different positions of the object O.

▶ In each case, the object O and its image I should be the same perpendicular distance from the mirror.

The reflected rays get further apart (diverge) and enter the eye. The eye follows the rays back in a straight line. The rays entering the eye appear to come from a point behind the mirror. This point is the image.

Note that the image in a plane mirror is not caused by real rays of light coming to a focus, as happens on a cinema screen. A mirror image is therefore called a virtual image.

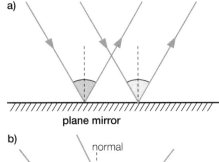

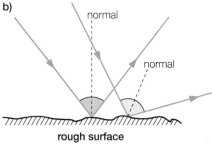

▲ **Figure 9.3** Light reflecting on **a)** a plane mirror and **b)** a rough surface

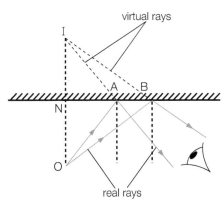

▲ **Figure 9.4** The image in a plane mirror

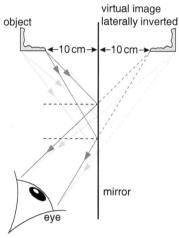

object   virtual image
          laterally inverted
←10 cm→←10 cm→
mirror
eye

▲ **Figure 9.5** This shows what happens when light from an object strikes a mirror

A mirror image is also reversed. If we hold a left-handed glove in front of a mirror, its image looks like a right-handed glove and vice versa (Figure 9.5). The image is said to be laterally inverted.

The image in a plane mirror is:

▶ virtual
▶ the same size as the object
▶ laterally inverted
▶ the same distance behind the mirror as the object is in front of the mirror.

Tip ↺

Be very precise when asked about the image size and position in a plane mirror.
Don't just write: Same size. Remember to add '... as the object'.
Don't just write: Same distance behind the mirror. Remember to add '... as the object is in front of the mirror'.

Test yourself ✎

1 State the size of the angle of incidence when the incident ray strikes a plane mirror at 90°.
2 The angle between a plane mirror and the incident ray is 40°. What size is the angle of reflection?
3 The angle between the incident ray and the reflected ray is 130°. What size is the angle of incidence?
4 Write the word 'AMBULANCE' in its laterally inverted form.
   Why is this laterally inverted form sometimes seen painted on real ambulances?
5 Two plane mirrors are inclined at right angles to each other. A ray of light strikes one mirror, M_1, at an angle of incidence of 30°, and the reflected ray from M_1 falls incident on M_2. Find the angle of reflection at M_2. Comment on the direction of the ray incident on M_1 and the reflected ray from M_2.
6 A student stands in front of a mirror and views his image. The student now takes a step backwards so that he is 20 cm further away from the mirror. By how much has the distance between the student and the image increased?
7 A letter L is placed in front of a mirror as shown in Figure 9.6.
   Copy the diagram and use the grid to draw the image of the letter L in the mirror.
8 Two mirrors A and B are arranged at 120° as shown in Figure 9.7. A ray of light is incident on mirror A.
   Calculate the angle of reflection of the ray reflected by mirror B.

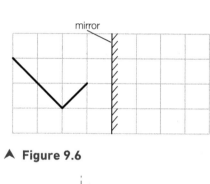

mirror

▲ **Figure 9.6**

A   65°   120°   B

▲ **Figure 9.7**

Refraction of light

Table 9.1

Material	Speed of light/m/s
Air (or vacuum)	3×10^8
Water	2.25×10^8
Glass	2×10^8

Refraction is the change in direction of a beam of light as it travels from one material to another due to a change in speed in the different materials. Table 9.1 shows the speed of light in various media. It is not necessary to remember the numbers in this table, but you must know that light travels faster in air than in water, and faster in water than in glass. You should also remember that the greater the change in the speed, the greater the angle it bends through (refracts).

▶ The angle between the normal and the incident ray is called the **angle of incidence**.

▶ The angle between the normal and the refracted ray is called the **angle of refraction**.

▶ The angle between the normal and the emergent ray is called the **angle of emergence**.

Figure 9.8 shows the paths of different light rays through a glass block. If the block has parallel sides, the angle of incidence is equal to the angle of emergence.

A ray parallel to the normal does not bend as it enters the block.

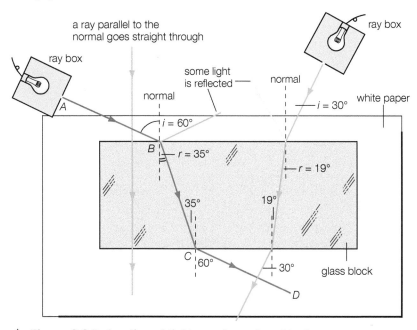

▲ **Figure 9.8** Refraction of light rays by a glass block

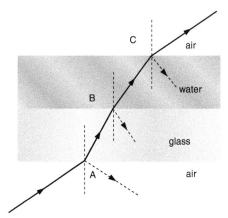

▲ **Figure 9.9** Diffraction through different media

Experiments show that:

▶ when light speeds up, it bends away from the normal

▶ when light slows down, it bends towards the normal.

Remember that this is also what happens to waves travelling from deep water into shallow water.

Figure 9.9 shows what happens when light travels from air through glass, then through water, and finally back into the air. Note the changes of direction in each case.

As the light enters the glass from the air at point A, it slows down, so it bends towards the normal.

As the light passes from glass into water at point B, it speeds up a little, so it bends away from the normal.

As the light passes from water into air at point C, it speeds up even more, so it bends still more away from the normal.

Using a ray tracing technique to measure angles of incidence and refraction as a ray of light passes from air into glass

Aims

▶ to trace rays through a glass (or perspex) block

▶ to measure angles of incidence and refraction

▶ to plot a graph of angle of incidence (*y*-axis) against angle of refraction (*x*-axis)

Apparatus

▶ rectangular glass (or perspex) block

▶ ray box

▶ low voltage power supply (PSU)

▶ leads

▶ protractor

▶ A4 plain white paper

▶ pencil

▶ ruler

Method

Figure 9.10 shows what we would expect to observe when light rays enter a glass block.

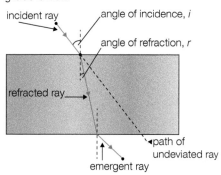

▲ **Figure 9.10** Light rays entering a glass block

1 Prepare a table for your results like the one shown on page 115.

2 Place the rectangular glass block in the centre of the sheet of white paper on a drawing board.

3 Draw around the outline of the block with a sharp pencil.

4 Connect the ray box to the PSU with the leads.

5 Switch on the PSU and direct a ray of light to enter the block near the middle of the longest side of the block, so that the angle of incidence is about 10°.

6 Mark the paths of both the incident ray and the emergent ray with two pencil dots, ensuring that the dots on each ray are as far apart as possible.

7 Carefully remove the glass block.

8 With a pencil and ruler, join the dots on the incident ray up to the point of incidence.

9 With a pencil and ruler, join the dots on the emergent ray back to the point of emergence.

10 With a pencil and ruler, draw a straight line between the point of incidence and the point of emergence to show the path of the refracted ray in the glass.

11 With a pencil and protractor, draw the normal at the point of incidence.

12 With the protractor, measure the angle of incidence, i, and the angle of refraction, r. Record the data in a table.

13 Repeat steps 1 to 12 for angles of incidence ranging from 20° up to 80°.

14 Observe that when the angle of incidence is zero (normal incidence), the angle of refraction is also zero.

Results

Angle of incidence, $i/°$	10	20	29	41	49	58	67	70
Angle of refraction, $r/°$	7	13	19	26	30	34	38	41

The above results are typical of those obtained in this experiment.

Treatment of the results

Plot a graph of angle of incidence (vertical axis) against angle of refraction (horizontal axis). Figure 9.11 shows an example of the type of graph that might be obtained.

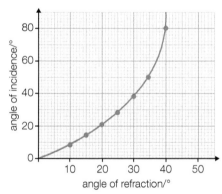

▲ **Figure 9.11** A graph showing angle of incidence against angle of refraction

Conclusion

The graph of angle of incidence against angle of refraction is a curve through the origin with an increasing gradient (the graph gets steeper as the angle of incidence increases). This tells us that i is not directly proportional to r (because the graph is not a straight line), but that i and r have a positive correlation (as i increases, r increases).

Tip

Always remember to put an arrow on real rays of light. Both the normal and the virtual rays are always dotted and never have an arrow.

Show you can

a) Describe what is meant by refraction of light.

b) Explain what happens to the speed of a ray of light if it refracts away from/towards the normal.

c) Describe an experiment to investigate the relationship between the angles of incidence and refraction when light is refracted as it passes from air into glass.

d) State what is meant by dispersion and state the conditions necessary for it to occur.

Dispersion of white light

All colours (frequencies) of light travel at the same speed in air, but different colours of light travel at different speeds in glass. This means that different colours bend by different amounts when they pass from air into glass. When light is passed through a triangular glass block (a prism), the effect is called dispersion and you can see a spectrum showing all the colours of the rainbow (Figure 9.12). Red light is bent (refracted) the least because it travels fastest in glass – it is slowed down the least. Violet light bends the most because it is slowest in glass – it is slowed down the most.

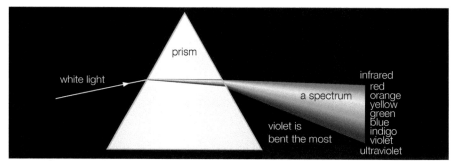

▲ **Figure 9.12** Dispersion of light through a prism

Test yourself

9 a) Figure 9.13 shows a ray of light passing from the air into the cornea of the eye.

    i) State the angle of incidence in air, and the angle of refraction in the cornea.

    ii) In which of the two media does light travel the fastest?

  b) What is the evidence for believing that red light travels faster in glass than blue light?

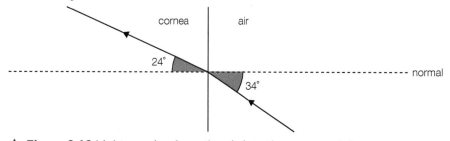

▲ **Figure 9.13** Light passing from the air into the cornea of the eye

10 A beam sound made underwater travels towards the surface. Sound travels faster in water than it does in air. Copy and complete Figure 9.14 to illustrate the refraction that occurs when the sound travels into the air. Mark the normal on your diagram.

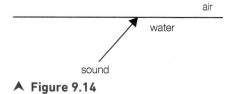

▲ **Figure 9.14**

11 Different shapes of glass prism are often used to change the direction of light rays.
Copy Figure 9.15 and continue the path of the ray shown until it emerges into the air.

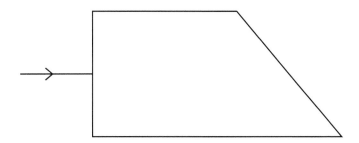

▲ **Figure 9.15**

12 Table 9.2 shows the speed of light in five different materials.
   a) Between which pair of materials does light show the greatest change in speed?
   b) Between which pair of materials does light refract the most?
   c) Does light bend towards or away from the normal when it passes from water into diamond? Give a reason for your answer.

Table 9.2

Material	Speed/m/s
Air	3.0×10^8
Ice	2.3×10^8
Water	2.2×10^8
Glass	2.0×10^8
Diamond	1.2×10^8

Critical angle

Figure 9.16 shows what happens when light travels through glass and emerges into the air. When the angle of incidence in glass is small enough, most of the light refracts into the air, but a little light is internally reflected (Figure 9.16a).

a)
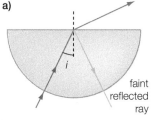
The angle of incidence is less than the critical angle.
Most of the light passes through into the air, but a little bit is internally reflected.

b)
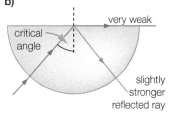
The angle of incidence is equal than the critical angle.
The emerging ray comes out along the surface. There is quite a bit of internal reflection.

c)
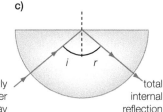
The angle of incidence is greater than the critical angle.
No light comes out It is all internally reflected, i.e. total internal reflection.

▲ **Figure 9.16** Total internal reflection of light

As the angle of incidence increases, the refracted ray bends closer to the glass and becomes weaker. At the same time, the light being internally reflected into the glass becomes stronger. Eventually, at a certain angle of incidence (called the critical angle), the light is refracted at an angle of 90° (Figure 9.16b). At this point, the refracted ray is very weak and the internally reflected ray is quite strong.

The critical angle for glass is about 42°. At angles above the critical angle, there is no refraction at all. All the light is reflected back into the glass – this is called total internal reflection (Figure 9.16c).

You should remember that:

▶ the critical angle is the angle of incidence in a material for which the angle of refraction in air is 90°

▶ at angles of incidence less than the critical angle, both reflection and refraction occur

▶ at angles of incidence greater than the critical angle, no refraction occurs, and the light is totally internally reflected.

You should investigate experimentally the critical angle and the conditions under which total internal reflection occurs within a semi-circular glass block. For total internal reflection to occur:

▶ the light must be travelling from glass (or perspex) towards the air

▶ the angle of incidence in the glass (or perspex) must be greater than the critical angle.

Practical activity

Experiment to find the critical angle

Apparatus

▷ semi-circular glass (or perspex) block
▷ ray box
▷ low voltage power supply (PSU)
▷ leads
▷ protractor
▷ A4 plain white paper
▷ pencil
▷ ruler

Method

1 Place the semi-circular glass block in the centre of the sheet of white paper on a drawing board.

2 Draw around the outline of the block with a sharp pencil.

3 Carefully remove the block and mark the centre, X, of the straight diameter.

4 Replace the glass block.

5 Connect the ray box to the PSU with the leads.

6 Switch on the PSU and direct a ray of light towards X, with light entering the glass from the air along the curved side as in Figure 9.17a. Ensure that the angle of incidence at the diameter is small so that light can be seen leaving the block at X.

7 Continue to direct the ray towards X, but slowly move the ray box so that the angle of incidence at X increases. Observe that the emergent ray at X becomes weaker and the internally reflected ray becomes stronger.

8 Continue to move the ray box to increase the angle of incidence at X until the refracted ray at X just emerges along the diameter, as in Figure 9.17b. Observe that if the angle of incidence at X is now increased even slightly, the light is totally internally reflected as in Figure 9.17c.

9 With a pencil, draw two dots on the incident ray as far apart as possible.

10 Carefully remove the glass block.

a)

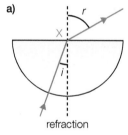

refraction

b)

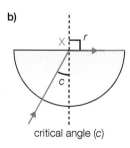

critical angle (c)

c)

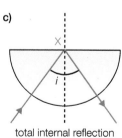

total internal reflection

▲ **Figure 9.17** Total internal reflection of light

11 With a pencil and ruler, join the dots on the incident ray beyond the point of incidence until the line reaches X.

12 With a pencil and protractor, draw the normal at X, the point of emergence.

13 With the protractor, measure the critical angle at point X and record the value in a table.

14 For reliability, repeat steps 1 to 13 about three more times.

15 Calculate the average of the values found for the critical angle.

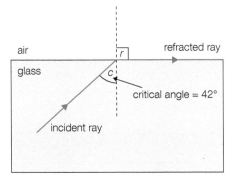

▲ **Figure 9.18** Illustrating the critical angle

Total internal reflection in parallel-sided glass blocks

When the angle of incidence in glass is equal to the critical angle, we just get refraction, as shown in Figure 9.18.

At angles of incidence above the critical angle, we get total internal reflection.

However, you can observe that in parallel-sided glass blocks, the incident and emergent rays are parallel. So a ray incident on one face can never be totally internally reflected at the opposite, parallel face. However, we can get total internal reflection if the ray is incident on the block from one side, as shown in Figure 9.19.

Total internal reflection in triangular prisms

Figure 9.20a shows how a ray of monochromatic (single colour) light is refracted by a triangular glass prism. Monochromatic light is used here to avoid unwanted dispersion.

Now look at what happens when light enters normally through one of the shorter sides, as in Figure 9.20b. Total internal reflection will occur once and the light emerges through the other short side after being bent through 90°.

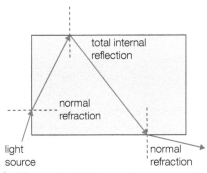

▲ **Figure 9.19** Illustrating the total internal reflection

Finally, consider what happens when light passes normally through the hypotenuse of a right-angled 45° prism as shown in Figure 9.20c. The angle of incidence on the first sloping face is 45°, which is just above the critical angle. So total internal reflection occurs. Similarly, total internal reflection will occur at the next sloping face, and the light eventually emerges parallel to the incident beam.

a)
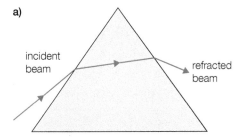

b)
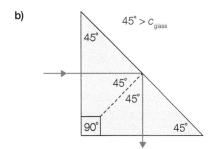

c)
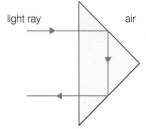

▲ **Figure 9.20** Refraction of monochromatic light in a triangular prism

Optical fibres

Optical fibres are lengths of solid glass core with an outer plastic sheath. Provided the fibre is not bent too tightly, light will strike the core–cladding boundary at an angle greater than the critical angle and be totally internally reflected at the surface of the glass core (Figure 9.21). However, every optical fibre has some imperfections at its reflecting surface and this means that the signal must be boosted every kilometre or so in communications links. Optical fibres are used to transmit both telephone and video signals over long distances.

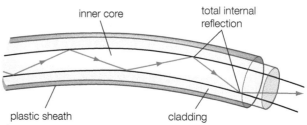

▲ **Figure 9.21** Passage of light through an optical fibre

The big advantage of optical fibres is that they can carry much, much more information than a copper cable of the same diameter. It has been estimated that the optical fibres in a sheath no thicker than a person's arm could carry all the telephone conversations taking place at any one time, all over the world.

What happens if the optical fibre is too tightly bent? If this happens, the angle of incidence at the core–cladding boundary may become less than the critical angle, and light will be lost by refraction into the cladding.

Endoscopes are used by surgeons to look inside a patient's body without needing to cut a large hole. They consist of bundles of optical fibres that allow light to travel into the body and allow image information to pass out of the body. The surgeon can therefore see on a monitor what is happening inside the body in real time. The endoscope kit also carries tools for cutting, snaring, water irrigation and retrieval of tissue. It is the use of optical fibres that makes keyhole surgery (laparoscopy) possible. Other uses of total internal reflection include prism binoculars and the prism periscope.

Show you can

a) Define what is meant by the term 'critical angle'.

b) State the two conditions required for total internal reflection to occur.

c) Describe in detail an experiment to measure the critical angle in a semi-circular glass block.

d) Describe the structure of an optical fibre and state some applications of fibre optics in telecommunications and medical endoscopy.

Test yourself

13 Write down a list of instructions you would give to a student as to how to measure the critical angle for glass using a semi-circular block.

14 A ray of light strikes an equilateral glass prism along the normal, as shown in Figure 9.22. The critical angle for the glass is 42°.

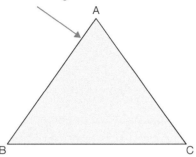

▲ **Figure 9.22** Ray of light striking an equilateral glass prism

a) State the angle of incidence in air at surface AB.

b) Calculate the size of the angle at A and hence the size of the angle of incidence at the surface AC.

c) State what happens to the light at surface AC.

d) Calculate the angle of refraction in air at surface BC.

e) Copy the diagram and complete it to show the passage of the light through the glass and back into the air.

15 Copy the diagrams from Figures 9.23 and 9.24 and the table. Copy and complete the table by placing ticks in the correct boxes.

	Is each statement below true or false?	True	False
![Figure 9.23] ▲ **Figure 9.23**	No refraction is taking place at the curved surface.		
	The angle of incidence in air at the curved surface is 0°.		
	The angle marked i is less than the critical angle.		

	Is each statement below true or false?	True	False
![Figure 9.24] ▲ **Figure 9.24**	No refraction is taking place at point X.		
	Refraction in the air will occur when the angle marked c is increased.		
	The light is travelling faster in the semi-circular block than it is in the air.		

16 Does light travelling in the core of an optical fibre travel faster, slower or at the same speed as light travelling in the cladding? Give a reason for your answer.

17 Explain why light can escape from the core of an optical fibre if it is bent too much.

Lenses

Lenses are specially shaped pieces of glass or plastic. There are two main types of lens (Figure 9.25), converging (or convex), and diverging (or concave).

converging (convex) lens

diverging (concave) lens

converging lens is thickest at the centre

diverging lens is thickest at the edges

▲ **Figure 9.25** The shapes of a converging lens and a diverging lens

There are two features of a converging lens that need to be defined:

▶ Rays of light parallel to the principal (central) axis of a converging lens all converge at the same point on the opposite side of the lens.

▶ This point lies on the principal axis, and is called the principal focus.

Figure 9.26 shows how a convex lens refracts light. Note that light refracts at each surface as it enters and leaves the lens, first bending towards the normal, and then away from the normal.

There are two features of a diverging lens (Figure 9.27) that need to be defined:

▶ Rays of light parallel to the principal axis of a diverging lens all appear to diverge from the same point after refraction in the lens.

▶ This point lies on the principal axis, and is called the principal focus.

The distance between the principal focus and the optical centre, C, of any lens is called the focal length.

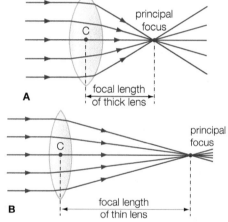

principal focus

focal length of thick lens

A

principal focus

focal length of thin lens

B

▲ **Figure 9.26** How light is refracted through a converging lens

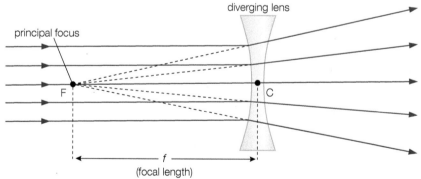

diverging lens

principal focus

F

C

f

(focal length)

▲ **Figure 9.27** How light is refracted through a concave lens

Note that light passing through the optical centre of a convex or concave lens is not bent. It passes straight through without refraction. Of course, light can pass through a lens from left to right or from right to left, so every lens has two principal foci and two focal lengths.

The only lenses you need to learn about at GCSE are equiconvex and equiconcave. This means that the principal foci on each side of the lens are the same distance from the optical centre. This becomes important when drawing ray diagrams to scale to find the position of an image.

Practical activity

This is a non-prescribed activity.

Aims
▶ to measure the focal length, f, of a converging lens using a distant object

Apparatus
▶ convex lens
▶ lens holder
▶ ruler
▶ sticky tape
▶ white screen in a holder
▶ distant object (such as a tree which can be seen through the windows in the laboratory and is at least 20 m away)

Method
1 Tape the ruler to the bench.
2 Place the white screen in its holder at the zero mark (see Figure 9.28).
3 Place the lens in its holder as close as possible to the screen.
4 Slowly move the lens away from the screen until the inverted image of the distant object is as sharp as possible.
5 Using the metre ruler, measure the distance from the centre of the lens to the white screen. This distance is the focal length of the lens, f.
6 Record the measured focal length in a prepared table.
7 For reliability, repeat steps 1 to 6 for four different distant objects, and determine the average value of f.

Diagram of the apparatus

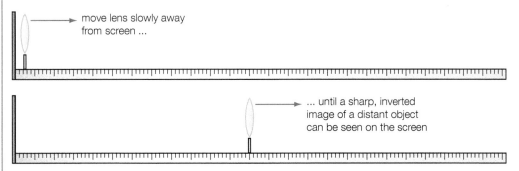

move lens slowly away from screen ...

... until a sharp, inverted image of a distant object can be seen on the screen

▲ **Figure 9.28** Apparatus used to measure the focal length of a converging lens

Results

Focal length f/mm	245	240	250	247	243
Average focal length f/mm	245				

(The above results are typical of those obtained using a lens of focal length around 250 mm.)

This is not an experiment where it is appropriate to write a conclusion.

Note that, regardless of the position of the object, the image in a concave (diverging) lens is always:
• erect
• virtual
• smaller than the object
• placed between the object and the lens.
This is shown in the ray diagram in Figure 9.29.

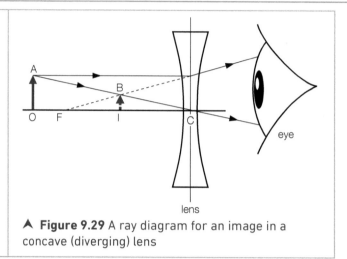

▲ **Figure 9.29** A ray diagram for an image in a concave (diverging) lens

Image in a convex (converging) lens

The position and properties of the image in a convex (converging) lens depends on the position of the object. We can find those positions and image properties by drawing a ray diagram.

Rules for drawing ray diagrams

To draw a ray diagram for a convex lens, you must draw at least two of the following rays:

▶ a ray parallel to the principal axis refracted through the principal focus on the other side of the lens

▶ a ray through the optical centre of the lens that does not change its direction (does not refract)

▶ a ray through the principal focus on one side of the lens which emerges so that it is parallel to the principal axis on the other side of the lens.

First steps when drawing a ray diagram for a convex lens:

▶ Using a ruler, draw a horizontal line to represent the principal axis and a vertical line for the lens.

▶ Mark the position of the principal focus with a letter F, the same distance from the optical centre on each side of the lens.

▶ Using a ruler, draw a vertical line touching the principal axis at the correct distance from the lens to represent the object.

▶ Using a ruler, draw at least two of the three construction rays, starting from the top of the object.

▶ Draw arrows on all rays to show the direction in which the light is travelling.

▶ The point where the construction rays meet is at the top of the image.

▶ The bottom of the image lies vertically below on the principal axis.

To illustrate the process, consider the following problem.

An object 5 cm tall is placed 6 cm away from a converging lens of focal length 4 cm. Find the position and height of the image.

In the solution shown in Figure 9.30, circled numbers have been added to show the order in which the lines or rays have been drawn. These numbers are drawn for illustration only and are normally omitted from such ray diagrams.

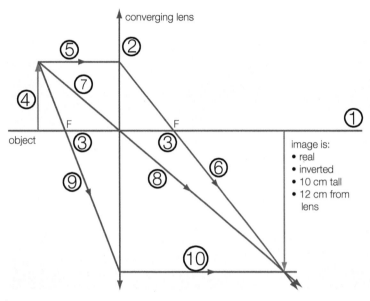

▲ **Figure 9.30** A ray diagram

The image is shown as a continuous line to show that it is real (can be projected on to a screen).

The downward arrow on the image shows it is inverted.

1 Horizontal line representing PA.

2 Vertical line representing lens.

3 Two principal foci marked F, each 4 cm from lens.

4 Object marked 6 cm from lens.

5 Ray from top of object parallel to PA …

6 … refracts through F.

7 Ray from top of object through the optical centre …

8 … is not refracted.

9 Ray from top of object through F …

10 … refracts parallel to PA.

Finally, the image is drawn from the point where the refracted rays meet to the PA.

Now use the ruler to measure the height of the image and its distance from the centre of the lens.

Ray diagrams

The ray diagrams in Figure 9.31 show where the image is formed for different positions of the object. You should carefully study the diagrams and Table 9.3, which gives a summary of the information.

a)

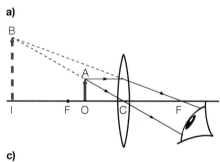

b)

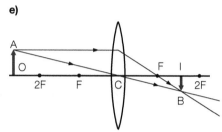

c)

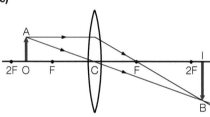

d)

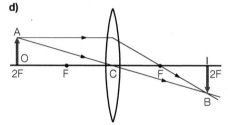

e)

▲ **Figure 9.31** Ray diagrams showing where the image is formed for different positions of the object

Table 9.3

Position of object	Location of image	Properties of image			
		Nature	Erect or inverted	Larger or smaller than object	Application
a) Between lens and F	On same side of lens as object, but further away from lens	Virtual	Erect	Larger	Magnifying glass
b) At F	At infinity	Real	Inverted	Larger	Searchlight
c) Between F and 2F	Beyond 2F	Real	Inverted	Larger	Cinema projector
d) At 2F	At 2F	Real	Inverted	Same size	Telescope – erecting lens
e) Just beyond 2F	Between F and 2F	Real	Inverted	Smaller	Camera
Very far away from lens	At F	Real	Inverted	Smaller	Camera

Show you can

a) Draw a ray diagram to show how a diverging lens produces a virtual, diminished image.

b) State the rules for drawing ray diagrams for a convex lens.

c) Draw a ray diagram to illustrate how a converging lens can produce:
   i) a real, magnified image
   ii) a diminished image
   iii) a virtual image.

d) State four properties of the image in a converging lens given the location of the object.

Tips ↻

You should pay particular attention to the ray diagrams which illustrate the principles of:
▶ the magnifying glass (to give an erect, virtual image)
▶ the projector (to give a magnified, real image)
▶ the camera (to give a diminished real image)
These diagrams are specifically required by the subject specification.

The human eye

Figure 9.32 is a diagram of the human eye. While most refraction occurs in the cornea, the purpose of the lens is to produce a sharp image on the light-sensitive tissue called the retina. Nerve endings here pass signals to the brain via the optic nerve.

In sophisticated cameras, the distance between the lens and the back of the camera can be changed to allow for focusing. In the eye this is done by muscles changing the shape, and hence the focal length, of the lens.

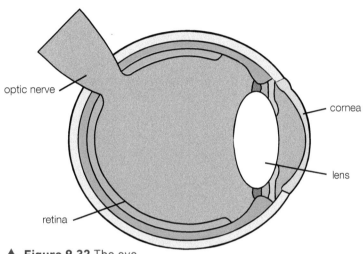

▲ **Figure 9.32** The eye

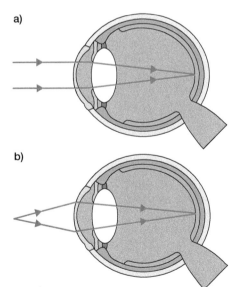

▲ **Figure 9.33** A ray diagram showing a) light from the far point reaching the eye and b) light from the near point reaching the eye

The normal eye

The farthest point that can be seen clearly by the eye is called the far point. For the normal eye, this is at infinity. Light from the far point reaches the eye as parallel rays (Figure 9.33a). The rays are refracted by the cornea and the eye lens so that they meet on the retina, where they form a sharp image.

The nearest point that can be seen clearly by the unaided eye without causing the muscles to strain is called the near point. For the normal eye, this is at 25 cm. The light from the near point reaches the eye as diverging rays (Figure 9.33b).

The cornea and the eye lens refract these rays so that they meet on the retina forming a sharp image.

a)

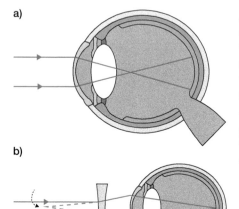

b)

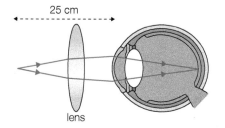

concave
lens

▲ **Figure 9.34** The use of a concave lens to assist an eye with myopia

A person who suffers from short sight (myopia) is unable to see distant objects sharply. They cannot make the lens thin enough to view distant objects. This causes the light from distant objects to converge just in front of the retina (Figure 9.34a). The image seen by the person is blurred.

To correct this defect a diverging lens is used. This means that parallel rays of light from a distant object are refracted so that they diverge before entering the eye (Figure 9.34b). The cornea and the eye lens can then bring these diverging rays to a sharp focus on the retina.

Long sight (hypermetropia)

A person with long sight sees distant objects clearly, but does not see near objects clearly. This happens because the eye muscles are too weak to make the lens thick enough for the light from an object at the near point (25 cm away) to form a sharp image on the retina (Figure 9.35a). The patient's near point is much further than 25 cm (Figure 9.35b).

a)

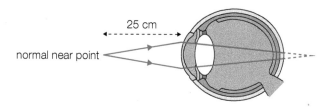

25 cm

normal near point

b)

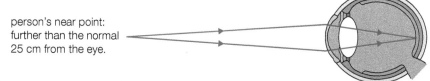

person's near point:
further than the normal
25 cm from the eye.

▲ **Figure 9.35** The near point of a normal and a hypermetropic eye

To correct for this defect, a converging lens is used. The light rays from an object at the near point are bent by this converging lens as shown in Figure 9.36. This light is now less diverging than before. The cornea and the eye lens can now bend the light so that a sharp image is formed on the retina.

25 cm

lens

▲ **Figure 9.36** The use of a converging lens to correct hypermetropia

1 State four properties of the image in a plane
mirror. *(4 marks)*

2 Two mirrors, M_1 and M_2, are placed at right
angles to one another. Figure 9.37 shows a ray of
light incident on mirror M_1 at an angle of 27° to
its surface.

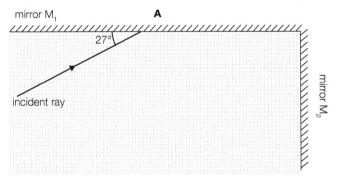

Figure 9.37

a) Copy the diagram onto graph paper and draw
the normal to mirror M_1 at point A. *(1 mark)*
b) Calculate the angles of incidence and
reflection at point A. *(2 marks)*
c) On your diagram, draw, as accurately as you
can, the reflected wave from A and from
mirror M_2. *(2 marks)*

3 Figure 9.38 shows a ray of light incident on a
glass block. Some of the light is reflected at the
top surface, and some of the light passes through
the glass and is reflected at the opposite side
which has a mirrored surface.

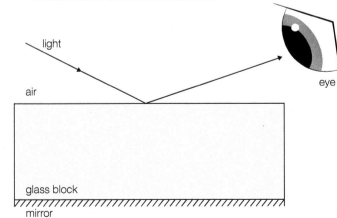

Figure 9.38

Copy the diagram and complete the path of the
ray of light through the glass block and back out
into the air towards the person viewing it. *(3 marks)*

4 Figure 9.39 shows the refraction of light in glass
and water.
a) Explain why the light bends as it enters both
glass and water. *(2 marks)*
b) How does Figure 9.39 suggest that the speed of
light in water is not the same as the speed
in glass? *(1 mark)*
c) Is light faster in glass or in water? Explain
your reasoning. *(2 marks)*

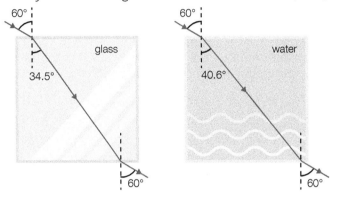

Figure 9.39

5 Figure 9.40 is a ray diagram that shows how a
lens can produce an image of an object. Each
small square corresponds to a distance of 2 mm.

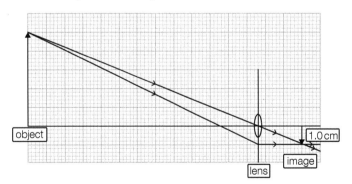

Figure 9.40

a) Which three of the words below best describes
the image? *(3 marks)*
real, diminished, erect, inverted, virtual, enlarged

b) What type of lens is shown in the ray
diagram? *(1 mark)*
c) Copy the diagram and mark on it the location
of the principal focus. *(1 mark)*
d) Find the focal length of the lens in
centimetres. *(1 mark)*

e) The magnification of the image is defined by the equation:

$$\text{magnification} = \frac{\text{height of image}}{\text{height of object}}$$

Use the equation to calculate the magnification of the image. *(1 mark)*

6 State the two conditions required for a ray of light to undergo total internal reflection as it moves from one substance to another. *(2 marks)*

7 A student was investigating how a ray of light passed through a semi-circular glass block. He drew the diagram shown in Figure 9.41 , but he made a number of mistakes.
Re-draw the diagram showing the correct paths of the two rays that were incorrectly drawn. *(2 marks)*

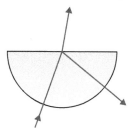

Figure 9.41

8 a) Explain what is meant by the statement that the critical angle for plastic is 42°. *(2 marks)*
  b) The red reflectors found on cars use total internal reflection to allow other road users to see the back of the vehicle. Figure 9.42 shows part of one such reflector. The red plastic has the shape of an isosceles right-angled triangle. Copy the diagram and complete the path of the ray shown through the red plastic and back into the air. *(4 marks)*

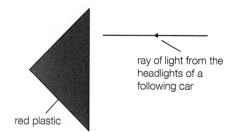

ray of light from the headlights of a following car

red plastic

Figure 9.42

9 Rainbows are a common sight when the Sun shines following a rain shower. Water droplets in the atmosphere are responsible for the colours seen. Look at Figure 9.43. At X, the sunlight is separated into many colours. The diagram shows only the red and violet light rays.

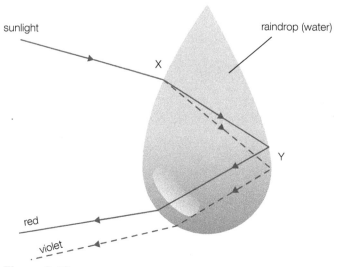

sunlight

raindrop (water)

X

Y

red

violet

Figure 9.43

a) What is this effect called? *(1 mark)*
b) State briefly why it happens. *(2 marks)*

10 Figure 9.44 shows what happens to parallel rays of light entering the eye of a patient.

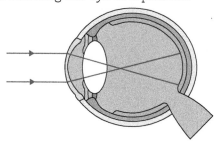

Figure 9.44

a) State the medical condition from which this patient is suffering. *(1 mark)*
b) What type of lens would an optician prescribe to correct this condition? *(1 mark)*
c) Draw a simple ray diagram to show how this type of lens would correct this condition. *(3 marks)*

11 A patient cannot clearly see an object which is about 25 cm from his eye. This is because the rays are brought to a virtual focus behind the retina (Figure 9.45).

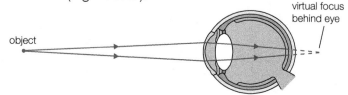

virtual focus behind eye

object

Figure 9.45

a) State the medical condition from which this patient is suffering. *(1 mark)*

b) What type of lens would an optician prescribe to correct this condition? *(1 mark)*

c) Draw a simple ray diagram to show how this type of lens would correct this condition. *(3 marks)*

12 A ray of light passes from air into water. The table shows how the angle of refraction in water changes for different angles of incidence in air.

Angle of refraction/°	0	15	29	41	48	49
Angle of incidence/°	0	20	40	60	80	90

a) Plot the graph of angle of refraction/° (*y*-axis) against angle of incidence/° (*x*-axis) and join the points with a smooth curve. *(5 marks)*

b) Does the graph show that the angle of refraction is directly proportional to the angle of incidence? Give a reason for your answer. *(1 mark)*

c) Use your graph to find:

   i) the angle of incidence in air when the angle of refraction in water is 35°. *(1 mark)*

   ii) the critical angle for light passing from water into the air. *(1 mark)*

10 Electricity

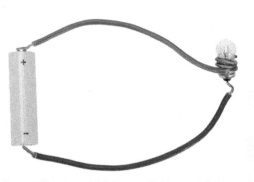

Specification points

This chapter covers specification points 2.3.1 to 2.3.22 of the GCSE Physics specification. It deals with conductors and insulators, simple circuits, symbols, charge, Ohm's law, lamps, series and parallel circuits and resistance.

Electricity

Electricity is an extremely versatile and useful form of energy. Many of our everyday activities depend on the use of electricity. It is hard to imagine life in today's society without it! Simple things such as entertainment, communications, transport and industry would simply grind to a halt if electricity ceased to exist.

When a woollen jumper is taken off over a nylon shirt in the dark, you can hear crackles and see tiny blue electric sparks. The nylon shirt has become charged with **static electricity**.

We now know that there are two types of charge, **positive** charge and **negative** charge. The negative charge is due to the presence of **electrons**, and it is only these particles that we are concerned with in this chapter.

When we connect a battery across a lamp, the lamp lights up. We say that the connecting wire (copper) and the filament of the bulb (tungsten) are both **electrical conductors**. The plastic covering is not an electrical conductor. We say the plastic is an insulator.

How can we tell if a material is a conductor or an insulator? To do that we connect the material in a circuit containing a battery, a bulb and the material being tested (Figure 10.1). If the bulb lights up, the material is a conductor. If the bulb does not light, the material is an insulator.

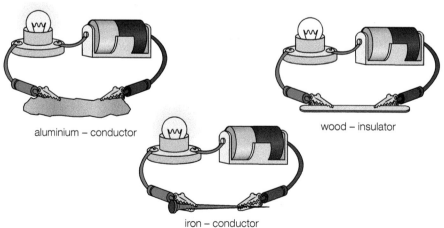

aluminium – conductor

wood – insulator

iron – conductor

▲ **Figure 10.1** Testing if a material is a conductor

In general, all metals are electrical conductors. Almost all non-metals are insulators, but there are a few exceptions. For example, graphite is a non-metal, but it conducts electricity.

Table 10.1 Common conductors and insulators

Good conductors	Gold	Silver	Copper	Aluminium	Mercury	Platinum	Graphite
Insulators	Polythene	Rubber	Wool	Wax	Glass	Paper	Wood

Why are metals good conductors?

An electric current is a flow of electrically charged particles. For the circuits we will consider, the charge involved is always the electron. Most electrons in atoms are bound by the positive nucleus to orbit in the surrounding shells. In metals, the outermost electron is often so weakly held that it can break away. We call such electrons free electrons. Some books call them delocalised electrons. Insulators contain no (or very few) free electrons.

So why does electrical current flow when we connect a metal wire between the terminals of a battery? An electric cell (commonly called a battery) can make electrons move, but only if there is a conductor connecting its two terminals to make a complete circuit. Chemical reactions inside the cell push electrons from the negative terminal round to the positive terminal. Figure 10.2 shows how an electric current would flow in a wire connected across a cell.

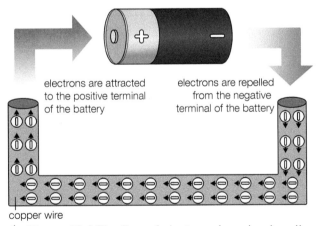

▲ Figure 10.2 The flow of electrons in a simple cell

Electrons are repelled from the negative terminal of the cell because like charges repel, and are attracted to the positive terminal because unlike charges attract. This make the electrons flow from the negative terminal to the positive terminal. Scientists in the nineteenth century thought that an electric current consisted of a flow of positive charge from the positive terminal of the cell to the negative terminal. This is now known to be incorrect, but we still refer to the direction of conventional current as flowing from positive to negative.

So, to summarise:

▶ Electrical conductors, like metals, have free electrons.
▶ Electrical insulators, like non-metals, have no free electrons.
▶ Free electrons move from the negative to the positive terminal of the battery.
▶ Conventional current is said to flow from the positive terminal to the negative terminal of the battery.

Standard symbols

An electrical circuit may be represented by a circuit diagram with symbols for components. Some widely used components are listed in Table 10.2. **Circuit diagrams** are easy to draw and are universally understood.

Table 10.2 Components of electrical circuits and their symbols

Component	Symbol	Appearance
switch		
cell	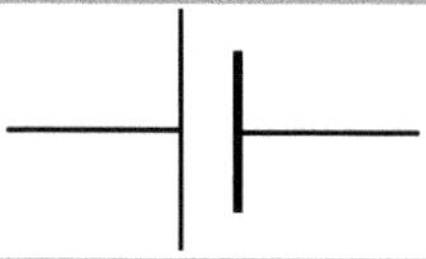	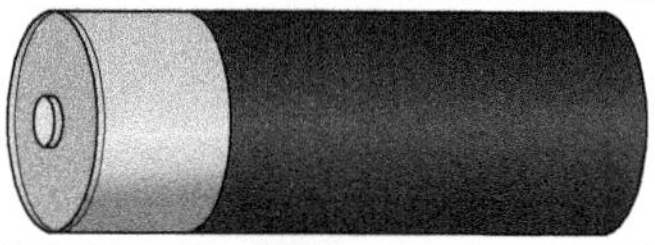
battery	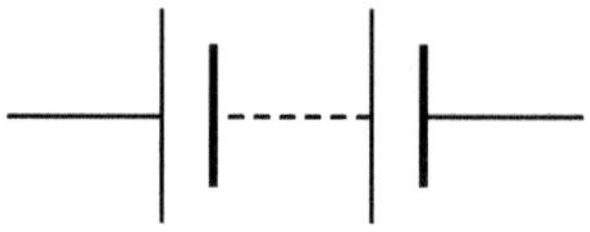	
resistor		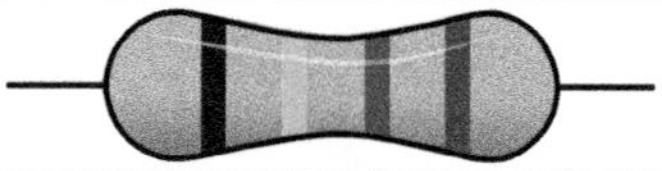
variable resistor (rheostat)	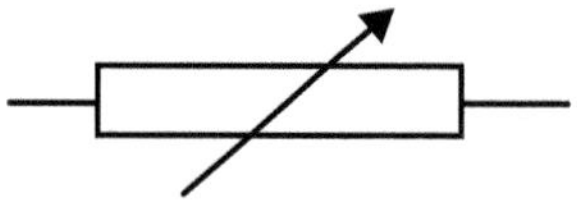	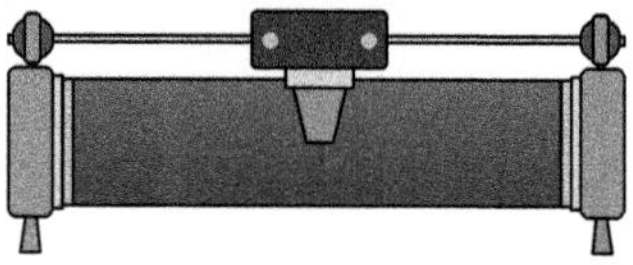
fuse		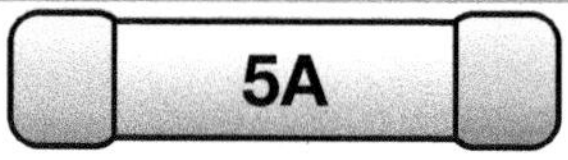
voltmeter	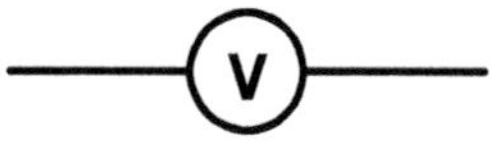	
ammeter	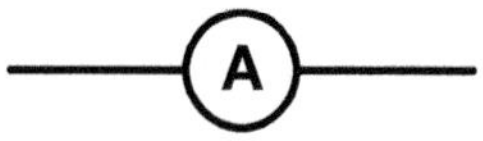	
lamp		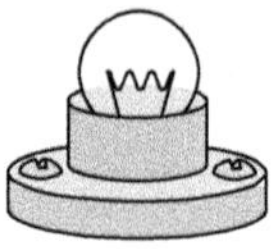

Cell polarity

By convention, the long, thin line in the symbol for a cell is taken as the positive terminal.

The short, bold line is the negative terminal.

Cells can be joined together minus to plus to make a **battery**. This leaves a positive and a negative terminal free to be connected into a circuit. Cells connected in this way are said to be connected in **series**.

Portable stereo systems have a number of cells connected in series. The reason for this is that the system requires a large voltage to operate. Connecting cells in series to make a battery increases the voltage. For example, connecting four 1.5 V cells in this way gives a 6 V battery, as shown in Figure 10.3.

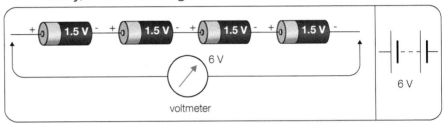

▲ **Figure 10.3** Cells correctly joined in series

You must be careful when connecting cells in series. If the **polarity** of one of the cells is reversed (in other words, if the positive and negative terminals of the cell are swapped) then the voltage is reduced dramatically. The voltages of two of the cells cancel each other out, leaving only one effective cell, as shown in Figure 10.4.

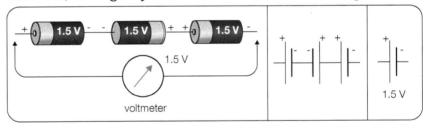

▲ **Figure 10.4** Cells incorrectly joined together

Notice that reversing the polarity of the cell in the middle has the effect of producing a battery of only 1.5 volts. The cells joined positive to positive cancel each other out.

The relationship between charge and current

The unit of charge is the **coulomb** (C). If it were possible to see a coulomb of charge, it would look like a very large assembly of electrons – about six million million million of them.

The unit of current is the **ampere** (A). Currents of around one ampere upwards can be measured by connecting an ammeter in the circuit. For smaller currents, a **milliammeter** is used. The unit in this case is the milliampere (mA). (1000 mA = 1 A). An even smaller unit of current is the **microampere** (μA). (1 000 000 μA = 1 A).

In general, if a steady current of I amperes flows for time t seconds, the charge Q coulombs passing any point is given by:

$$Q = I \times t$$
$$\text{charge} = \text{current} \times \text{time}$$

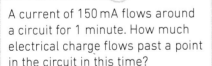

Example

A current of 150 mA flows around a circuit for 1 minute. How much electrical charge flows past a point in the circuit in this time?

Answer

$I = 150 \, \text{mA}$

$\quad = 0.15 \, \text{A}$

$t = 1 \, \text{minute}$

$\quad = 60 \, \text{s}$

$Q = I \times t$

$\quad = 0.15 \times 60$

$\quad = 9 \, \text{C}$

Notice that:

▶ charge is measured in coulombs (C) and given the symbol Q
▶ current is measured in amperes (A) and given the symbol I
▶ time is measured in seconds (s) and given the symbol t.

If a charge of 1 C passes a fixed point in 1 s, a current of 1 A is flowing in the circuit.

There are two conditions which must be met before an electric current will flow:

▶ There must be a complete circuit – i.e. there must be no gaps in the circuit.
▶ There must be a source of energy so that the charge may move – this source of energy may be a cell, a battery or the mains power supply.

Show you can

a) Explain the difference between conductors and insulators in terms of free electrons.
b) Determine the direction of electron flow and conventional current given a circuit diagram.
c) Know the standard symbols for electrical components and be able to use them to draw simple circuit diagrams.
d) Know what is meant by cell polarity and calculate the voltage of a battery with regards to cell polarity.

Test yourself

1 The cells in Figure 10.5 are all identical. The total battery voltage is 1.6 V.

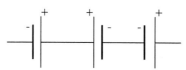

▲ **Figure 10.5** Three battery cells in series

a) Calculate the voltage of each cell.
b) Redraw the battery showing the connections which would give a battery capable of delivering the maximum possible voltage.
c) State the maximum voltage which this battery could deliver.

2 Convert the following currents into milliamperes:
a) 3.0 A
b) 0.2 A
c) 200 μA

3 Convert the following currents into amperes:
a) 400 mA
b) 1500 mA
c) 500 000 μA

4 What charge is delivered if:
a) a current of 6 A flows for 10 seconds?
b) a current of 300 mA flows for 1 minute?
c) a current of 500 μA flows for 1 hour?

5 Calculate the currents that flow when the following charges pass a fixed point in the following times:
a) 100 C, time = 5 s
b) 500 mC, time = 50 s
c) 60 mC, time = 200 s

Resistance

The resistance R of an electrical conductor can be found using the ammeter–voltmeter method. We measure the current I through the conductor when a voltage V is applied across its ends. The resistance R is then calculated using the equation:

$$R = \frac{V}{I}$$

where: R is the resistance in ohms (Ω)

V is the voltage in volts (V)

I is the current in amperes (A)

This is the basis of the Ohm's law experiment which follows.

Prescribed practical

Ohm's law

During your course, you are required to carry out a number of prescribed practical experiments. This is one of them.

Aims

▶ to pass an electric current through a wire

▶ to measure the current for different values of the voltage across the wire

▶ to take precautions to ensure the temperature of the wire is kept constant

▶ to plot a graph of voltage across the wire (*y*-axis) against current in the wire (*x*-axis)

▶ to use the graph to establish an equation linking voltage and current

▶ to determine the resistance of the wire

Apparatus

▶ low voltage power supply unit (PSU)

▶ rheostat

▶ ammeter

▶ voltmeter

▶ connecting leads

▶ resistance wire

▶ switch

Method

1 Prepare a table for your results like that shown on the next page.

2 Ensure that the PSU is switched off and connect it to the mains socket.

3 Set up the circuit as shown in Figure 10.6. The device marked R represents the wire being tested.

4 Adjust the PSU to supply zero volts.

5 Switch on the PSU.

6 Record the voltage on the voltmeter and the corresponding current on the ammeter.

7 Switch off the PSU immediately after recording in the table values for voltage and current.

8 Wait for about two minutes to ensure the wire cools to room temperature.

9 Switch on the PSU and adjust the voltage (or the rheostat) so that the reading on the voltmeter increases by 1.0 V.

10 Repeat steps 6–9 until readings have been recorded for a voltages ranging from zero to a maximum voltage of 6 V*. This is Trial 1.

11 Repeat the entire experiment to obtain a second set of values for current. This is Trial 2.

Calculate the mean current from the two trials and enter the results in the table.

Plot the graph of voltage against mean current.

* It is necessary to ensure the wire's temperature remains constant (close to room temperature).

We do this by:

▶ keeping the voltage low (so that the current remains small)

▶ switching off the current between readings (to allow the wire to cool).

Table for results
Table 10.3

Voltage/V	0.00	1.00	2.00	3.00	4.00	5.00	6.00
(Trial 1) Current/A							
(Trial 2) Current/A							
Mean current/A							
Ratio of voltage to current/Ω							

Circuit diagram

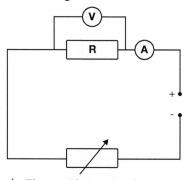

▲ **Figure 10.6** A circuit diagram

Treatment of the results

Plot the graph of voltage/V (vertical axis) against mean current/A (horizontal axis). It should look similar to the graph in Figure 10.7.

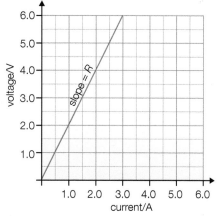

▲ **Figure 10.7** A graph showing voltage against current

Discussion of the results

The graph of V against I is a straight line through the origin. This tells us that the current in a metallic conductor is directly proportional to the voltage across its ends, provided the temperature remains constant.

This result is commonly called **Ohm's law**.

Measuring the resistance

The resistance of the wire does not change when the current and voltage change. The resistance of a wire at constant temperature depends only on three factors:

▶ the material from which the wire is made

▶ the length of the wire

▶ the cross-sectional area of the wire.

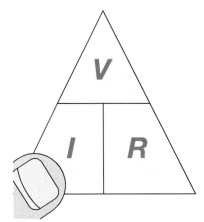

▲ Figure 10.8 The resistance formula triangle

Tip

Most mathematical questions in physics start with an equation, so it is essential that you learn your formulae!

The graph of V against I is a straight line through the origin. The ratio of V to I is constant throughout the experiment. It also means that, in this case, the slope of the graph of V against I is equal to the resistance of the wire. However, you should understand that measuring the slope of the V–I graph is not, in general, the correct way to measure a resistance. This will become clearer in the next experiment (see page 140).

Earlier, we established the resistance formula:

$$V = I \times R$$
$$\text{voltage} = \text{current} \times \text{resistance}$$

Figure 10.8 may help you to 'change the subject' in the resistance formula. To find the equation for current, put your thumb over I and you can see that $I = \dfrac{V}{R}$.

Similarly to find R, put your thumb over it and it is clear that $R = \dfrac{V}{I}$.

Example

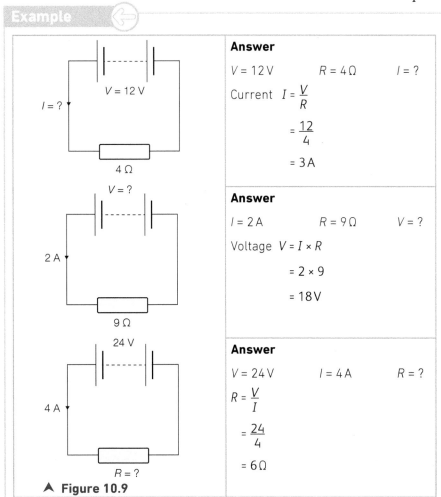

▲ Figure 10.9

Answer

$V = 12\,V$ $R = 4\,\Omega$ $I = ?$

Current $I = \dfrac{V}{R}$

$= \dfrac{12}{4}$

$= 3\,A$

Answer

$I = 2\,A$ $R = 9\,\Omega$ $V = ?$

Voltage $V = I \times R$

$= 2 \times 9$

$= 18\,V$

Answer

$V = 24\,V$ $I = 4\,A$ $R = ?$

$R = \dfrac{V}{I}$

$= \dfrac{24}{4}$

$= 6\,\Omega$

Show you can

a) Explain why it is necessary to keep the temperature of the resistance wire constant in the Ohm's law experiment, and describe how this is achieved.

b) Explain why it is necessary to obtain several values of current and voltage in the Ohm's law experiment, and describe how this is achieved.

c) Describe the *V–I* characteristic graph for a metal wire at constant temperature, and state what conclusion can be drawn from it.

d) State Ohm's law and the condition under which it is valid.

Test yourself

6 Calculate the current flowing through a 10 Ω resistor which has a voltage of 20 V across it.

7 A resistor has a voltage of 15 V across it when a current of 3 A flows through it.
Calculate the resistance of the resistor.

8 A current of 2 A flows through a 25 Ω resistor. Find the voltage across the resistor.

9 A voltage of 15 V is needed to make a current of 2.5 A flow through a wire.
a) What is the resistance of the wire?
b) What voltage is needed to make a current of 2.0 A flow through the wire?

10 There is a voltage of 6.0 V across the ends of a wire of resistance 12 Ω.
a) What is the current in the wire?
b) What voltage is needed to make a current of 1.5 A flow through it?

11 A resistor has a voltage of 6 V applied across it and the current flowing through is 100 mA. Calculate the resistance of the resistor.

12 A current of 600 mA flows through a metal wire when the voltage across its ends is 3 V. What current flows through the same wire when the voltage across its ends is 2.5 V?

Practical activity

You should note that this is **not** a prescribed practical activity and will not be assessed in Unit 3A.

Resistance of a filament lamp

Aims

▶ to pass an electric current through a filament lamp

▶ to measure the current for different values of the voltage across the lamp

▶ to plot a graph of voltage across the lamp (*y*-axis) against current in the lamp (*x*-axis)

▶ to use the graph to determine the resistance of the lamp at two different values of current

Variables

The dependent variable is the current flowing in the lamp.

The independent variable is the voltage across the lamp.

The controlled variables are the length of the filament (inside the lamp) and its cross-sectional area.

Apparatus

▶ low voltage power supply unit (PSU)

▶ rheostat

▶ ammeter

▶ voltmeter

▶ connecting leads

▶ filament lamp in a suitable holder

▶ switch

Method

1 Prepare a table for your results like that shown in Table 10.4 on the next page.

2 Ensure that the PSU is switched off, and connect it to the mains socket.

3   Set up the circuit as shown in Figure 10.10.

4   Adjust the PSU to supply zero volts.

5   Switch on the PSU.

6   Record the voltage on the voltmeter and the corresponding current on the ammeter.

7   Adjust the voltage (or the rheostat) so that the reading on the voltmeter increases by 0.5 V.

8   Repeat steps 6–7 until readings have been recorded for voltages ranging from zero to a maximum voltage of 6 V (for a 6 V lamp). This is trial 1.

9   Repeat the entire experiment to obtain a second set of values for current. This is trial 2.

10 Calculate the mean current from the two trials and enter the results in the table.

11 Plot a graph of voltage against mean current. Your graph should look similar to the graph in Figure 10.11.

Table for results

Table 10.4

Voltage/V	0.00	1.00	2.00	3.00	4.00	5.00	6.00
(Trial 1) Current/A							
(Trial 2) Current/A							
Mean current/A							
Ratio of voltage to current/Ω							

Circuit diagram

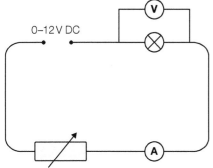

▲ **Figure 10.10**

Treatment of the results

Plot the graph of voltage/V (vertical axis) against mean current/A (horizontal axis).

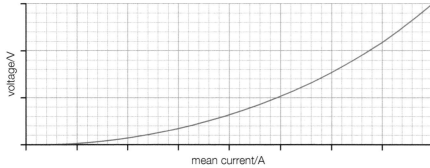

▲ **Figure 10.11**

Discussion of the results

Notice that we tried hard to keep the temperature constant in the previous experiment. This is because the resistance of a metal can rise rapidly when the temperature increases. In this experiment, the temperature of the filament is allowed to rise – at first the lamp only glows orange, but as the current rises, it becomes white hot at its operating temperature. As its temperature rises, so does the resistance.

Measuring the resistance

You should now find the resistance of the filament at two points on your curve. You can do this by calculating the ratio of the voltage to the current for the two different values of the current. You will notice that the resistance rises with increasing current.

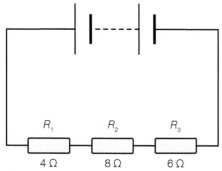

▲ **Figure 10.12a** Calculating the total resistance of three resistors in a series circuit

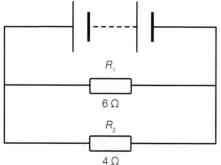

▲ **Figure 10.12b** Calculating the total resistance of two resistors in parallel

Series circuits

The total resistance of two or more resistors in series is simply the sum of the individual resistances of the resistors.

$$R_{total} = R_1 + R_2 + R_3$$

In Figure 10.12a, the three resistors could be replaced by a single resistor of $4 + 8 + 6 = 18\,\Omega$.

Parallel circuits

Figure 10.12b shows a parallel circuit with two resistors. The total resistance of two equal resistors in **parallel** is half of the resistance of one of them. The total resistance of three equal resistors in parallel is one third of the resistance of one of them, and so on.

When considering **two** unequal resistances, R_1 and R_2, in parallel, we use the 'product over sum' formula:

$$R_{total} = \frac{R_1 \times R_2}{R_1 + R_2}$$

$$= \frac{product}{sum}$$

The formula for calculating the total resistance of three resistors in parallel is:

$$\frac{1}{R_{total}} = \frac{1}{R_1} + \frac{1}{R_2} + \frac{1}{R_3}$$

This last formula can be extended to any number of resistors in parallel.

Example

1 Find the combined resistance of two $8\,\Omega$ resistors
   a) in series and
   b) in parallel.

Answer
   a) For resistors in series: $R_{total} = R_1 + R_2$

   $= 8 + 8$

   $= 16\,\Omega$

   b) The total resistance of two equal resistors in parallel is half of the resistance of one of them, so in this case, the total resistance is $8 \div 2 = 4\,\Omega$.

2 A 6 Ω resistor and a 3 Ω resistor are connected
   a) in series and
   b) in parallel.
   In each case, find the resistance of the combination.

Answer

a) For resistors in series: $R_{total} = R_1 + R_2$

$$= 6 + 3$$

$$= 9 \, \Omega$$

b) For two unequal resistors in parallel, we use the product over sum rule. Note that this only works for two resistors.

$$R_{total} = \frac{\text{product}}{\text{sum}}$$

$$= \frac{R_1 \times R_2}{R_1 + R_2}$$

$$= \frac{6 \times 3}{6 + 3}$$

$$= 2 \, \Omega$$

3 A 24 Ω resistor, a 12 Ω resistor and a 8 Ω resistor are connected
   a) in series and
   b) in parallel.
   In each case, find the resistance of the combination.

Answer

a) For resistors in series: $R_{total} = R_1 + R_2 + R_3$

$$= 24 + 12 + 8$$

$$= 44 \, \Omega$$

b) For three unequal resistors in parallel, we use:

$$\frac{1}{R_{total}} = \frac{1}{R_1} + \frac{1}{R_2} + \frac{1}{R_3}$$

$$= \frac{1}{24} + \frac{1}{12} + \frac{1}{8}$$

$$= \frac{1}{4}$$

So, $R_{total} = 4 \, \Omega$

The final example illustrates how to cope with a combination of resistors in parallel with resistors in series.

4 Find the total resistance of the combination shown in Figure 10.13.

▲ **Figure 10.13**

Answer

Using the product over sum formula, we see that the 4 Ω and 6 Ω parallel combination gives a total resistance of 2.4 Ω. Similarly, the 9 Ω and 18 Ω parallel combination gives a total resistance of 6.0 Ω.

There is, therefore, a series combination of 2.4 Ω + 1.6 Ω + 6.0 Ω, which gives a total resistance of 10 Ω.

Current and voltage in series circuits

When resistors are connected in series:

▶ the current in each resistor is the same
▶ the sum of the voltages across each resistor is equal to the battery or power supply voltage.

Example

1 Resistances of 2 Ω, 4 Ω, and 6 Ω are connected in series across a 3 V battery. Calculate:
   a) the total resistance of the circuit
   b) the current in each resistor
   c) the voltage across each resistor.

Comment on your answer to part c).

Answer

a) $R_{total} = R_1 + R_2 + R_3$
$$= 2 + 4 + 6$$
$$= 12 \, \Omega$$

This means that the circuit is equivalent to one with 12 Ω placed across a 3 V battery.

b) $I = \dfrac{V}{R}$
$$= \dfrac{3}{12}$$
$$= 0.25 \, A$$

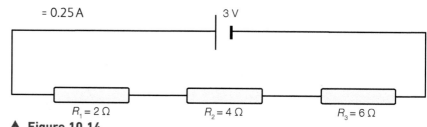

▲ **Figure 10.14**

c) 2Ω $V = I \times R$ 4Ω $V = I \times R$ 2Ω $V = I \times R$
       $= 0.25 \times 2$ $= 0.25 \times 4$ $= 0.25 \times 6$
       $= 0.5 \, V$ $= 1.0 \, V$ $= 1.5 \, V$

Comments

The sum of the voltages (0.5 V + 1.0 V + 1.5 V) is 3 V, which is exactly the same as the voltage of the battery.

The voltages are in exactly the same proportion as the resistances. So, the 4 Ω resistor has twice the voltage as the 2 Ω resistor and the 6 Ω resistor has three times the voltage as the 2 Ω resistor.

Voltage in parallel circuits

When resistors are connected in parallel:

▶ the voltage across each resistor is the same as the voltage provided by the battery
▶ the sum of the currents in each resistor is equal to the current coming from the battery.

1 Resistances of $2\,\Omega$ and $3\,\Omega$ are connected in parallel across a $6\,V$ battery (see Figure 10.15).

a) State the voltage across each resistor.

b) Calculate:

i) the total resistance of the circuit

ii) the current in each resistor

iii) the total current taken from the battery.

Answer

a) $6\,V$

b) i) $R_{total} = \dfrac{\text{product}}{\text{sum}}$

$= \dfrac{R_1 \times R_2}{R_1 + R_2}$

$= \dfrac{2 \times 3}{2 + 3}$

$= 1.2\,\Omega$

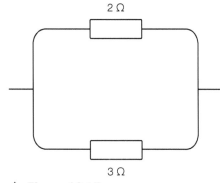

▲ **Figure 10.15**

ii) $2\,\Omega \quad I = \dfrac{V}{R} \qquad 3\,\Omega \quad I = \dfrac{V}{R}$

$\qquad\quad = \dfrac{6}{2} \qquad\qquad\quad = \dfrac{6}{3}$

$\qquad\quad = 3\,A \qquad\qquad\quad = 2\,A$

iii) We can do this in two different ways:

$I_{total} = \dfrac{V_{battery}}{R_{battery}} \qquad$ or $\quad I_{battery}$ = sum of currents in resistors

$\qquad = \dfrac{6}{1.2} \qquad\qquad\qquad = 3 + 2$

$\qquad = 5\,A \qquad\qquad\qquad\quad = 5\,A$

2 Resistances of $4\,\Omega$, $6\,\Omega$ and $12\,\Omega$ are connected in parallel across a battery. A current of $2.0\,A$ flows from the battery towards the parallel network (see Figure 10.16).

Calculate

a) the total resistance of the network

b) the battery voltage and the voltage across each resistor

c) the current in each resistor.

Comment on the answers to parts a) and b).

Answer

a) $\dfrac{1}{R_{total}} = \dfrac{1}{R_1} + \dfrac{1}{R_2} + \dfrac{1}{R_3}$

$= \dfrac{1}{12} + \dfrac{1}{6} + \dfrac{1}{4}$

$= \dfrac{1}{2}$

So $R_{total} = 2\,\Omega$

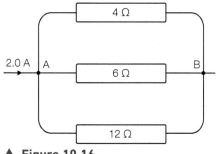

▲ **Figure 10.16**

b) $V_{battery} = I_{battery} \times R_{total}$

$\qquad = 2.0 \times 2$

$\qquad = 4\,V$

voltage across each resistor = 4 V

c) $4\,\Omega \quad I = \dfrac{V}{R}$ $\qquad 6\,\Omega \quad I = \dfrac{V}{R}$ $\qquad 12\,\Omega \quad I = \dfrac{V}{R}$

$\qquad\quad = \dfrac{4}{4}$ $\qquad\qquad = \dfrac{4}{6}$ $\qquad\qquad\quad = \dfrac{4}{12}$

$\qquad\quad = 1\,A$ $\qquad\qquad = 0.67\,A$ $\qquad\qquad\quad = 0.33\,A$

Comments

In b), the voltages across each resistor are the same as the battery voltage.

The sum of the currents in the parallel network (1.0 A + 0.67 A + 0.33 A) is 2.0 A, which is exactly the same as the current from the battery.

The currents in each resistor are in inverse proportion to the resistance. So the current in the 6 Ω resistor is twice the current in the 12 Ω resistor, and the current in the 4 Ω resistor is 3 times the current in the 12 Ω resistor.

Hybrid circuits

Hybrid circuits contain both series and parallel elements. We treat each part separately, eventually finding the total resistance of the entire network, as shown in the example.

Example

A parallel combination of 2 Ω and 3 Ω resistors is joined in series with a 5 Ω resistor. The network is connected across a 31 V battery. Calculate the current taken from the battery, the voltage across each resistor, and the current in each resistor.

Answer

From Example 1 on page 145, we see that the combined resistance of the two resistors in parallel is 1.2 Ω. So the total resistance of the circuit, R_{total}, is

$5 + 1.2 = 6.2\,\Omega$

The current taken from the battery:

$I_{battery} = \dfrac{V_{battery}}{R_{battery}}$

$\qquad = \dfrac{31}{6.2}$

$\qquad = 5\,A$

The current in the 5 Ω resistor is 5 A.

The voltage across the 5 Ω resistor is:

$V = I \times R$

$\quad = 5 \times 5$

$\quad = 25\,V$

Since the battery voltage is 31 V, the voltage across the parallel network is:

$31 - 5 = 6\,V$

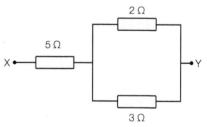

▲ **Figure 10.17**

So the voltage across the 2 Ω resistor = voltage across the 3 Ω resistor = 6 V

current in 2 Ω $= \dfrac{V}{R}$ current in 3 Ω $= \dfrac{V}{R}$

$= \dfrac{6}{2}$ $= \dfrac{6}{3}$

$= 3\,A$ $= 2\,A$

Test yourself

13 a) Calculate the value of the current from the cell in each of these circuits.

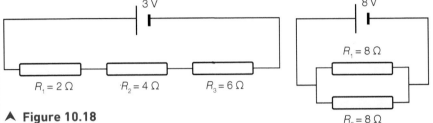

▲ **Figure 10.18**

b) State the voltage across each 8 Ω resistor in the second circuit.

14 a) In each of the following circuits, calculate the currents I_1, I_2 and I_3 shown by ammeters A_1, A_2 and A_3.

b) Calculate the voltage across each resistor.

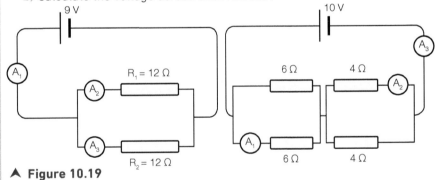

▲ **Figure 10.19**

15 a) In circuits A and B below, all the lamps are identical. Copy and complete the table that follows.

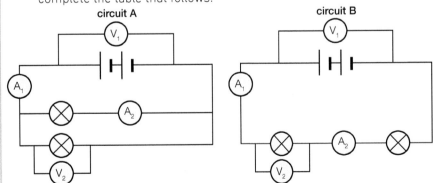

▲ **Figure 10.20**

circuit	V_1/V	V_2/V	I_1/A	I_2/A
A	3	3	0.4	
B	3		0.1	

b) Calculate the resistance of each lamp.

Resistance and length

Earlier in this chapter, we stated that the resistance of a piece of wire depended on its length. You are required to carry out a prescribed experiment to show the nature of this dependence.

Prescribed practical

Resistance and length

Aims
▶ to pass an electric current through a wire whose length can be varied
▶ to measure the current and voltage for different lengths of wire
▶ to take precautions to ensure the temperature of the wire is kept constant
▶ to calculate the resistance for each length of wire
▶ to plot a graph of resistance of the wire (*y*-axis) against length of the wire (*x*-axis)
▶ to use the graph to establish an equation linking resistance and length

Variables
▶ The independent variable is the length of the wire.
▶ The dependent variable is the resistance of the wire.
▶ The controlled variables are the temperature and the cross-sectional area of the wire.

Apparatus
▶ low voltage power supply unit (PSU)
▶ rheostat
▶ ammeter
▶ voltmeter
▶ connecting leads
▶ resistance wire
▶ switch
▶ metre stick
▶ sticky tape

Method
1 Prepare a table for your results like that shown on the next page.
2 Measure and cut off one metre of nichrome resistance wire.
3 Attach it with sticky tape to a metre ruler – make sure there are no kinks in the wire.
4 Set up the circuit as shown in Figure 10.21.
5 Ensure that the PSU is switched off and connect it to the mains socket.
6 Adjust the PSU to supply about 1 V.
7 Connect the 'flying lead' so that the length of wire across the voltmeter is 10 cm.
8 Switch on the PSU.
9 Record the voltage on the voltmeter and the corresponding current on the ammeter.
10 Switch off the PSU immediately after recording in the table values for voltage and current.
11 Wait for about 2 minutes to ensure the wire cools to room temperature*.
12 Switch on the PSU again.

13 Repeat steps 7–13 until readings have been recorded for lengths of wire ranging from 10 cm to 90 cm.

14 Calculate the resistance of each length of wire, using $R = \dfrac{V}{I}$.

15 Plot the graph of resistance (*y*-axis) against length (*x*-axis).

* It is necessary to ensure the wire's temperature remains constant (close to room temperature).

We do this by:

▶ keeping the voltage low (so that the current remains small)

▶ switching off the current between readings to allow the wire to cool.

Table for results

Length of wire/cm	10	20	30	40	50	60	70	80	90
Voltage across wire/V									
Current in wire/A									
Resistance of wire/Ω									

Circuit diagram

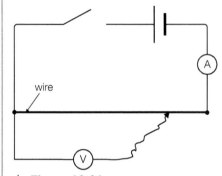

▲ **Figure 10.21**

Treatment of the results

Plot the graph of resistance/Ω (vertical axis) against length/cm (horizontal axis). Your graph should look something like that shown in Figure 10.22.

Discussion of the results

The graph of resistance/Ω against length/cm is a straight line through the origin. This tells us that the resistance of a metal wire is directly proportional to its length, provided its cross-sectional area and the temperature remain constant. This means there is a mathematical relationship between the resistance, *R*, and the length, *L*.

The relationship is:

$R = kL$

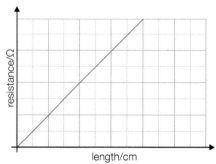

▲ **Figure 10.22**

where *k* is the gradient of the graph.

Since $k = \dfrac{R}{L}$, the unit for *k* is Ω/cm (or Ω/m).

Note that the value of *k* depends on the material of the wire and its cross-sectional area.

A reel of constantan wire of length 250 cm has a total resistance of 15.0 Ω
Calculate:

a) the resistance of 1.0 m of wire
b) the length of wire needed to have a resistance of 3 Ω
c) the resistance of a 90 cm length of the wire.

Answer.

a) $k = \dfrac{R}{L}$

$= \dfrac{15\,\Omega}{250\,cm}$

$= 0.06\,\Omega/cm$

$= 6\,\Omega/m$, so the resistance of 1.0 m of wire is 6 Ω

b) $L = \dfrac{R}{k}$

$= \dfrac{3\,\Omega}{0.06\,\Omega/cm}$

$= 50\,cm$

c) $R = kL$

$= 6\,\Omega/m \times 0.9\,m$

$= 5.4\,\Omega$

Practical activity

You should note that this is **not** a prescribed practical activity.

Resistance and cross-sectional area

Aims

▶ to pass an electric current through wires of constant length but different cross-sectional area

▶ to measure the current for different values of the voltage and hence find the resistance

▶ to measure the cross-sectional area of each wire

▶ to plot graphs of resistance (*y*-axis) against cross-sectional area (*x*-axis), and resistance (*y*-axis) against 1/area (*x*-axis)

▶ to determine the relationship between resistance and area

Variables

The dependent variable is the resistance of each wire.

The independent variable is the cross-sectional area of each wire.

The controlled variables are the length of the wire and the material from which it is made.

Apparatus

▶ low voltage power supply unit (PSU)

▶ rheostat

- ammeter
- voltmeter
- connecting leads
- filament lamp in a suitable holder
- switch
- wooden dowel or pencil or micrometer

Method

1 Prepare seven samples of constantan wire, all 50 cm long and all with different cross-sectional areas.

2 Prepare a table for your results like that shown below.

3 As a preliminary, use the micrometer screw gauge (Figure 10.23c) to measure the diameter, d, of one of the wires, or measure the length (L) of 20 turns of a resistance wire wound tightly together on a pencil or wooden dowel (Figure 10.23b) and divide this length by 20 to calculate its diameter.

4 Calculate the cross-sectional area, A, using $A = \frac{\pi d^2}{4}$ and record the data in the table of results.

5 Repeat this process for six further thicknesses of the same length of wire and same type of material.

6 Ensure that the PSU is switched off and connect it to the mains socket.

7 Set up the circuit as shown in Figure 10.23a.

8 Switch on the PSU and adjust it if necessary to obtain a voltage of 2 V.

9 Record the voltage on the voltmeter and the corresponding current on the ammeter.

10 Determine the resistance of this specimen of wire using $R = \frac{V}{I}$ and record the data in the table.

11 Repeat steps 7–10 for each of the other wire specimens.

12 Plot the graph of resistance against cross-sectional area and resistance against $\frac{1}{area}$.

Table for results

Table 10.6

Mean diameter/mm							
Cross-sectional area/mm²							
Voltage/V							
Cross-sectional area/mm²							
Resistance/Ω							
$\frac{1}{area}$ /1/mm²							

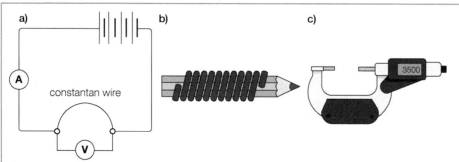

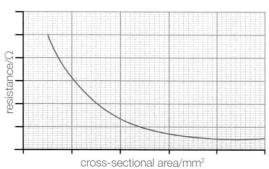

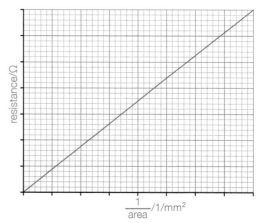

▲ **Figure 10.23 a)** Circuit diagram **b)** Wire wrapped around pencil **c)** Micrometer

Treatment of the results

Plot the graphs of:

a) resistance/Ω against cross-sectional area/mm² (which should be similar to the graph in Figure 10.24)

b) resistance/Ω against $\frac{1}{\text{area}}$ /1/mm² (which should be similar to the graph shown in Figure 10.25)

▲ **Figure 10.24** Graph of resistance against cross-sectional area

▲ **Figure 10.25** Graph of resistance against 1/area

The graph of R against A is a curve with decreasing gradient.

The graph of R against $\frac{1}{A}$ is a straight line through the origin. This tells us that the resistance is inversely proportional to the cross-sectional area.

This means there is a mathematical relationship between the resistance, R, and the cross-sectional area, A.

The relationship is: $R = \frac{k}{A}$

where k is the gradient of the straight line graph of R against $\frac{1}{A}$.

Since $k = RA$, the unit for k is $\Omega\,cm^2$ (or $\Omega\,m^2$ or $\Omega\,mm^2$).

Note that, for a given wire, the value of k depends on the material of the wire and its length.

Example

A length of constantan wire has a resistance of 15.0 Ω and a cross-sectional area of 0.3 mm². Calculate:

a) the resistance of the same length of constantan wire of area 0.9 mm².

b) the cross-sectional area of the same length of constantan wire if its resistance is 9 Ω.

Answer

a) $k = RA$

$= 15\,\Omega \times 0.3\,mm^2$

$= 4.5\,\Omega\,mm^2$

$R = \dfrac{k}{A}$

$= \dfrac{4.5\,\Omega\,mm^2}{0.9\,mm^2}$

$= 5\,\Omega$

b) $A = \dfrac{k}{R}$

$= \dfrac{4.5\,\Omega\,mm^2}{9\,\Omega}$

$= 0.5\,mm^2$

Show you can

a) Describe how the resistance of a piece of metal wire depends on:
   i) its length
   ii) its cross-sectional area and
   iii) the material from which it is made.

b) Describe in detail the experiments which demonstrate how the resistance of a piece of metal wire depends on
   i) its length
   ii) its cross-sectional area and
   iii) the material from which it is made.

Practical activity

Investigating how the resistance of a metallic conductor at constant temperature depends on the material it is made from

The controlled variables in this investigation are the length and the thickness of wire.

It is reasonable to expect that the resistance should depend on the type of material from which a wire is made.

Using one metre of 32 swg (standard wire gauge) copper wire, measure and record the resistance as before.

Then repeat the process using the same dimensions of wires such as manganin, nichrome, constantan and copper.

When comparing wires of the same length and cross-sectional area, you should find that the order of increasing resistance is: copper, manganin, constantan and nichrome.

Test yourself

16 A school buys a reel of constantan wire. The supplier's data sheet says that the wire has a resistance of 2.5 Ω/m. Calculate:
   a) the length of wire a technician must cut from the reel to give a resistance of 2 Ω and
   b) the resistance of a 120 cm length of wire cut from the reel.

17 When a current of 0.15 A flows through a 48 cm length of eureka wire the voltage across its ends is 0.90 V. What length of the same type of wire would give a current of 0.36 A when the voltage across its ends is 1.44 V?

18 An 80 cm length of wire A has a resistance of 2.4 Ω. The resistance of a 50 cm length of wire B is 1.2 Ω. A student cuts a 30 cm length, L_1, of wire from a reel of wire A and a 40 cm length L_2 from a reel of wire B.
   a) Which length, L_1 or L_2, has the greater resistance? Explain your reasoning.
   b) Sketch a graph of resistance against length for wire A and wire B, and state which graph has the larger gradient.

19 A technician cuts an 80 cm length of wire from a reel marked 3.0 Ω/m. The technician joins the two free ends of the wire together to form a loop. She then attaches two crocodile clips to the wire at opposite ends of a diameter.
   a) Explain why the total resistance between the crocodile clips is 0.6 Ω.
   b) In what way, if at all, does the total resistance between the crocodile clips change, if one of the clips is moved along the wire towards the other? Explain your reasoning.

20 A 50 cm length of wire, of diameter 0.2 mm has a resistance of 1.6 Ω. Find the resistance of:
   a) a 75 cm length of wire of the same material and same diameter
   b) a 50 cm length of wire of the same material and diameter 0.4 mm
   c) a 75 cm length of wire of the same material and diameter 0.4 mm.

21 A piece of wire that is 20 cm long, has a diameter of 0.3 mm and a resistance of 0.8 Ω.
   Show that the resistance of a wire of the same material that has a length of 80 cm and diameter 0.6 mm, is also 0.8 Ω.

1

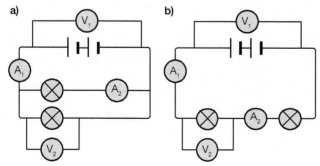

Figure 10.26

a) What flows in the direction indicated by the arrow in Figure 10.26? *(1 mark)*

b) Copy the circuit diagram and mark on it an arrow to show the direction in which charged particles flow through the two resistors. *(1 mark)*

c) What name is given to these charged particles? *(1 mark)*

d) The current in the larger resistor is 0.6 A. State the size of the current in the smaller resistor. *(1 mark)*

e) Show that the electrical charge delivered by the battery in one minute is 36 C. *(3 marks)*

2 Resistors of 20 W and 10 W are connected in series across a battery. The current in the larger resistor is 0.6 A. Show that the battery voltage is 18 V. *(3 marks)*

3 a) Two identical resistors are connected in parallel across a battery. Their combined resistance is 12 Ω. What is the resistance of each resistor? *(1 mark)*

b) A different pair of identical resistors are connected in series across a battery. Their combined resistance is also 12 Ω. What is the resistance of each resistor? *(1 mark)*

4 In Figure 10.27 a) and b), the lamps are identical and each has an internal resistance of 6 Ω. Each cell in the battery can supply a voltage of 1.5 V. For each circuit, find the reading which would be shown on the four meters. *(4 marks)*

a)   b)

Figure 10.27

5 Four identical resistors are arranged as shown in Figure 10.28.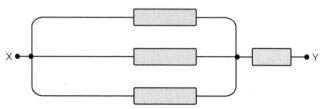

Figure 10.28

The current entering at X is 3 mA, and the voltage between X and Y is 12 mV. Calculate:

a) the total resistance between X and Y *(3 marks)*

b) the current in each resistor *(2 marks)*

c) the voltage across each resistor *(2 marks)*

d) the resistance of each resistor. *(2 marks)*

6 In Figure 10.29, resistors R_1 and R_2 have resistances of 40 Ω and 20 Ω respectively.

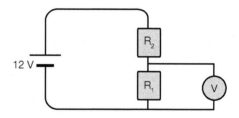

Figure 10.29

a) Calculate the voltage you would expect to observe on the voltmeter. *(3 marks)*

b) What assumption have you made about the resistance of the voltmeter itself? *(2 marks)*

7 Look at the circuit in Figure 10.30. Copy and complete Table 10.7 to show the effective resistance between X and Y for the four different switch settings. *(4 marks)*

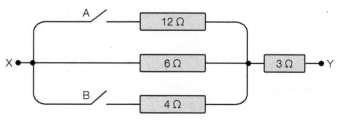

Figure 10.30

Table 10.7

Switch		Effective resistance between X and Y/Ω
A	B	
open	open	
open	closed	
closed	open	
closed	closed	

8 A student investigating the relationship between the resistance and the cross-sectional area of a resistance wire of length 1.00 m obtains the following results. The material from which the wire is made is always the same.

a) Copy the table and enter the missing data.

(3 marks)

Area of cross section of wire/mm²	0.5	1.0	2.0	3.0	4.0
Resistance/Ω	24.0		6.0		

b) Calculate the resistance of 50 cm of a wire of cross-sectional area 1.5 mm² made from the same material.

(3 marks)

Specification points

This chapter covers specification points 2.3.23 to 2.3.30 of the GCSE Physics specification. It deals with electrical energy, power and electricity in the home.

In the previous chapter, we saw that electricity generates heat when it passes through a metal wire. Why does that happen?

As electrons pass through the conductor they collide with the atoms. In these collisions the light electrons lose energy and the heavy atoms gain energy. This causes the atoms to vibrate faster. Faster vibrations mean a higher temperature. This is called joule heating.

This heating effect is put to good use in devices such as hairdryers and toasters (Figure 11.1), which contain heating elements usually made from nichrome wire.

Electrical energy

If 1 coulomb of charge gains or loses 1 joule of energy between two points, there is a voltage of 1 volt between those two points.

$$\text{voltage} = \frac{\text{energy transferred}}{\text{charge}}$$

Rearranging this formula gives us:

energy transferred = voltage × charge

It is much easier to measure current than charge.

We know that:

charge = current × time

Substituting for charge gives a very useful formula for **energy transferred** in an electric circuit:

$$\text{energy transferred} = \text{voltage} \times \text{current} \times \text{time}$$
$$E = V \times I \times t$$

Electrical power

Earlier, you learned that the formula for mechanical power of a machine is defined as the rate at which energy is transferred and is given by:

$$\text{power} = \frac{\text{energy transferred}}{\text{time}}$$

We have seen that in an electrical circuit:

energy transferred = voltage × current × time

F

▲ **Figure 11.1** Toasters and hairdryers use an electrical current to heat

Substituting for energy transferred:

$$\text{power} = \frac{\text{voltage} \times \text{current} \times \text{time}}{\text{time}}$$

or

electrical power = voltage × current

This is often expressed by the equation $P = I \times V$.

Domestic appliances such as toasters, hairdryers and TVs have a power rating marked on them in watts or in kilowatts (1 kW = 1000 W).

Some widely used electrical devices and their typical power ratings are shown in Figure 11.2.

Example

1 If 0.5 A flows through a bulb connected across a 6 V power supply for 10 seconds, how much energy is transferred?

Answer

energy = $V \times I \times t$

$\qquad$ = 6 × 0.5 × 10

$\qquad$ = 30 J

2 A study lamp is rated at 60 W, 240 V. How much current flows in the bulb?

Answer

power = $V \times I$

$\quad$ 60 = 240 × I

$\quad$ $I = \dfrac{60}{240}$

$\qquad$ = 0.25 A

Show you can

a) Explain, in terms of particle collisions, why an electrical current flowing in a resistance wire generates heat.

b) State the equation relating power, voltage and current.

200 W

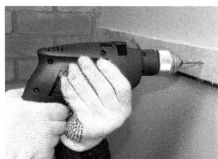

20 W

1100 W

2200 W

60 W

▲ **Figure 11.2** A variety of electrical devices found in the home

F

Test yourself

1 How much electrical energy does a 1000 W convector heater consume in one hour?
2 In 10 seconds, an electric toaster consumes 15 000 joules of energy from the mains supply. What is its power:
   a) in watts?
   b) in kilowatts?
3 A study lamp draws a current of 0.25 A at 240 V from the mains supply. Calculate:
   a) the power and
   b) the amount of energy it consumes in 60 seconds.
4 a) The starter motor of a car has a power rating of 960 W. If it is switched on for 5 seconds, how much energy does it use?
   b) The same starter motor is powered by connecting it to a 12 V car battery. How much current does it use?

Example

What power is dissipated in a 10 Ω resistor when the current through it is:

a) 2 A
b) 4 A?

Answer

a) power = I^2R

    = $2^2 \times 10$

    = 40 W

b) power = I^2R

    = $4^2 \times 10$

    = 160 W

The three equations for power

We can use the equations for power and voltage to express power in three different ways.

$$P = I \times V \qquad\qquad V = I \times R$$
$$P = IV$$
$$= I^2R$$
$$P = \frac{V^2}{R}$$

This formula (known as Joule's law) can be used to calculate the electrical power **dissipated** (converted) into heat in resistors and heating elements. The heat dissipated is sometimes referred to as the 'ohmic losses'.

The example opposite shows that when the current is doubled, the power dissipated is quadrupled!

This idea has important implications for electricity transmission in the next chapter.

Test yourself

5 Find the power of an electric heater if it takes a current of 4 A when connected to a 240 V supply.
6 Find the current in a light bulb of power 48 W when connected to a 12 V supply.
7 Copy and complete the table for domestic appliances, all of which operate at 240 V.

Name of appliance	Power rating	Current drawn	Resistance
Bulb of study lamp	60 W		
Television	80 W		
Toaster	1200 W		
Convector heater	2 kW		
Shower	3 kW		

8 Calculate how much power an electric kettle uses if it has an element of resistance 48 Ω and is connected to the 240 V mains supply.

Example

If electricity costs 16 pence per kilowatt hour, find the number of units used by a 2800 W oven when it is switched on for five hours. Calculate the cost of using this oven for that time.

Answer

number of units used
= power rating × time
     (kW)   (hour)

= 2.8 kW × 5 h

= 14 kWh

total cost = number of units used ×
         cost per unit

     = 14 kWh × 16 pence

     = 224 pence

     = £2.24

Paying for electricity

Electricity companies bill customers for electrical energy in units known as kilowatt hours (kWh). These are sometimes called 'units' of electricity.

One kilowatt hour is the amount of energy transferred when 1000 W is delivered for one hour. You should be able to prove for yourself that:

1 kWh = 3 600 000 J = 3.6 MJ

The following two formulae are very useful in calculating the cost of using a particular appliance for a given amount of time:

number of units used = power rating (in kilowatts) × time (in hours)

total cost = number of units used × cost per unit

Figure 11.3 shows an example of an electricity bill.

Northern Electricity Board				Customer account no: 3427 364
Present meter reading	Previous meter reading	Units used	Cost per unit (incl. VAT)	£
57139	55652	1487	15.0p	£223.05

▲ **Figure 11.3** An electricity bill

The difference between the current reading and the previous reading is the number of units used.

In this particular example, 57 139 − 55 652 = 1487 units (kWh).

1487 units have been used.

If the cost of a unit is known, then the total cost of the electricity used can be determined.

In this particular example: 1487 units at 15.0 pence per unit = 22 305 pence or £223.05.

One-way switches

This kind of switch acts as a make-or-break device to switch a circuit on or off (Figure 11.4). When the switch is open, there is air between the conducting contacts. Since air is an insulator, the circuit is incomplete. No current flows.

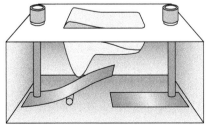
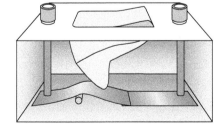

▲ **Figure 11.4** A one way switch

The rocker (the part of the switch that you press) is made of plastic. This is important with high voltages to prevent current flowing through the body of the user. When the rocker is pressed, the conducting contacts are pushed together. There is now a complete circuit, so current can flow.

As we will see later, it is important that the switch is placed in the live side of a circuit.

Two-way switches

In most two-storey houses, you can turn the landing lights on or off from upstairs or downstairs. Two-way switches are used for this.

Figure 11.5a illustrates a two-way switch circuit in one of its two 'on' positions.

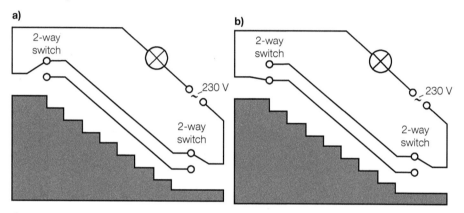

▲ **Figure 11.5** A two-way switch circuit

When both switches are up (or both are down), the circuit is complete and a current flows through the bulb.

At the top of the stairs, one of the switches is pressed down and the circuit is broken, as shown in Figure 11.5b. The bulb goes dark.

Two-way switches may be used in a number of different situations, such as to control a light from opposite ends of a long corridor.

Figure 11.6 shows the 4 possible states of a two-way switch used in a simple circuit with a battery.

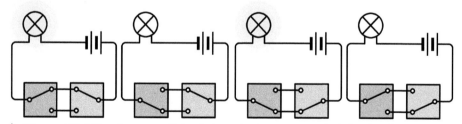

▲ **Figure 11.6** The four possible states of a two-way switch in circuit

How to wire a three-pin plug

▶ The wire with the blue insulation is the neutral wire – connect this to the left-hand pin.

▶ The brown insulated wire is the live wire – connect this to the right-hand pin.

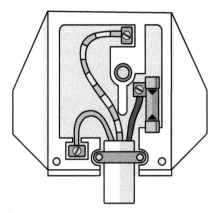

Figure 11.7 How to wire a three-pin plug

▶ The wire with the yellow and green insulation is the earth wire – connect this to the top pin.
▶ Each of these wires should be wrapped around its securing screw so that they are tightened as the screw is turned.
▶ Insert the correct cartridge fuse into its holder beside the live wire.
▶ Finally, fix the 3-core cable tightly with the cable grip and screw on the plug-top.

Each pin in the plug fits into a corresponding hole in the socket. The earth pin is longer than the others so that it goes into the socket first and pushes aside safety covers, which cover the rear of the neutral and live holes in the socket.

Fuses

A fuse is a device which is meant to prevent damage to an appliance.

The most commonly used fuses are either a 3 A (red) fuse for appliances up to 720 W, or 13 A (brown) fuse for appliances between 720 W and 3 kW.

If a larger-than-usual current flows, the fuse will melt and break the circuit.

Nowadays, residual current circuit breakers (RCCBs) are becoming much more common. They work by detecting any difference between the currents in the live and neutral wires. When a difference is detected due to a fault they break the circuit very rapidly before there is any danger. Unlike fuses, which are designed to protect the appliance, RCCBs protect both the appliance and the user because they are very sensitive and very quick.

Selecting a fuse

Every appliance has a power rating. How much current the appliance will use is found using the power formula:

power = voltage × current

For example, a jig-saw has a power of 350 W. The current it draws when connected to the mains is given by:

$$current = \frac{power}{voltage}$$
$$= \frac{350}{240}$$
$$= 1.46 A$$

This is the normal current the device uses. A larger current could destroy it.

A 3 A fuse would allow a normal working current to flow and protect the jig-saw from larger currents. A 13 A fuse would allow a dangerously high current to flow without breaking the circuit. It is important to use the correct size of fuse.

Remember that a fuse protects the appliance. It does **not** protect the person using the appliance. It can take 1 to 2 seconds for a fuse to melt – enough time for the user to receive a fatal electric shock.

If an appliance becomes live, a current flows through the earth wire and then from the socket earth connection to the earth via a substantial metal connection such as a water pipe. During this process, the fuse in the plug will blow. Before the fuse in a faulty appliance is replaced, the appliance should be checked by a qualified electrician.

The earth wire

An **earth wire** can prevent harm to the user.

Suppose someone using an electric drill accidentally drilled into a mains electricity wire hidden inside a wall (Figure 11.8). The mains current would flow through the drill bit and into the metal casing of the drill. The casing would become live, and if someone were to touch it they would get a possibly fatal electric shock as the current rushes through their body to earth. The earth wire can help to prevent this – it offers a low resistance route of escape, enabling the current to go to earth by a wire rather than through a human body. Because of this low resistance, the current through the fuse will be large. After a few seconds, the large current will melt the fuse.

Together, the fuse and the earth wire help to reduce the risk of an electric shock.

Any appliance with a metal casing could become live if a fault developed, so such appliances nearly always have a fuse fitted in the plug.

Double insulation

Appliances such as vacuum cleaners and hairdryers are usually double insulated. The appliance is encased in an insulating plastic case and is connected to the supply by a two-core insulated cable containing only a live and a neutral wire. Any metal attachments that a user might touch are fitted into the plastic case so that they do not make a direct connection with the motor or other internal electrical parts. The symbol for double insulated appliances is shown in Figure 11.9.

The live wire

Earlier, we said that the switch and fuse must be placed on the live side of the appliance (Figure 11.10). Why is this important? The live wire in a mains supply is at high voltage (effectively around 230 V). The neutral side is at approximately zero volts.

If a fault occurs and the fuse blows, the live, dangerous wire is disconnected. If the fuse were on the neutral side, the appliance would still be live, even when the fuse had blown.

Switches are also placed on the live side for the same reason. If the switch were on the neutral side, the appliance would still be live, even when the switch was in the OFF position.

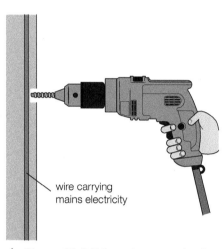

▲ **Figure 11.8** If there is no earth wire connected to the casing of the drill, the current will flow through the user

wire carrying mains electricity

▲ **Figure 11.9** The symbol for double insulation

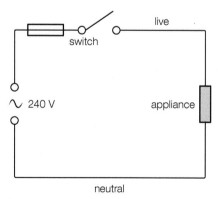

switch
live
240 V
appliance
neutral

▲ **Figure 11.10** This fuse has been correctly placed on the live side of the power supply

F

1 An electric fire is connected to the mains supply by means of a three-pin plug.
  a) The electric fire has a rating of 2000 W when used on a 230 V mains supply. Calculate which fuse would be most suitable for use with this fire: 1 A, 3 A, 5 A or 13 A. *(4 marks)*
  b) Describe the size of the current in the live, neutral and earth wires when the fire is switched on and working properly. *(1 mark)*
  c) The live wire becomes loose and comes into contact with the metal body of the electric fire. Describe the danger that could arise when the electric fire is switched on. *(1 mark)*
  d) How should the earth wire be connected to reduce this danger? *(1 mark)*
  e) How should the fuse be connected so as to reduce this danger? *(1 mark)*
  f) Explain how the action of the earth wire and the fuse reduce this danger. *(1 mark)*

2 a) Mrs Johnston's electricity meter was read at 1 month intervals.
  Reading on 1 April 2017   11 897 kW h
  Reading on 1 May 2017   12 107 kW h
  i) How many units of electricity were used in the Johnston home during April? *(1 mark)*
  ii) If 1 unit of electricity costs 15p, calculate the cost of electricity to Mrs Johnston during April. Show clearly how you get your answer. *(2 marks)*
  b) An electric fire is wired using a three-pin plug as shown in Figure 11.11.

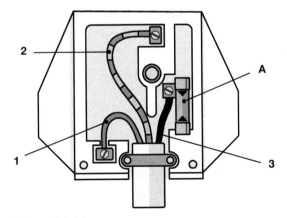

Figure 11.11
  i) Name the part labelled A. *(1 mark)*
  ii) Which of the wires 1, 2 or 3 should be connected to the metal casing of the fire? *(2 marks)*
  iii) State the colours of the insulation on wires 1, 2 and 3. *(3 marks)*

3 a) A television set is marked 240 V, 80 W.
  i) Explain carefully what these numbers mean. *(2 marks)*
  ii) The flex that connects a brand new television to the mains has only two wires inside it. An electrician confirms that there should only be two wires inside the plug. Explain why only two wires are needed. *(2 marks)*
  iii) To which of these wires should the switch on the television be connected? *(1 mark)*
  iv) Apart from allowing the user to switch the television on and off, the switch is connected in this way for another reason. What is this other reason? *(1 mark)*
  v) Explain how the owner of this television is protected from possible electric shock. *(1 mark)*
  b) An electric oven is rated at 8 kW.
  i) Calculate the cost of using the oven to cook for 2 hours. The cost of electricity is 15p per unit. Show clearly how you get your answer. *(3 marks)*
  ii) When the oven is on, the same current passes through the cable as the heating elements. Explain why the cable does not heat up. *(2 marks)*

4 Calculate how much electrical energy, in kilowatt hours, is used for:
  a) a 100 W lamp on for 12 hours *(2 marks)*
  b) a 250 W television on for 4 hours *(2 marks)*
  c) a 2400 W kettle on for 5 minutes. *(2 marks)*

5 An electric shower is rated at 230 V, 15 A.
  a) Calculate the electrical power used by the shower heater. *(2 marks)*
  b) Calculate the cost of a 10-minute shower if 1 kWh costs 12p. *(2 marks)*

6 a) Calculate the amount of electrical energy, in joules, used by a 1000 W electric fire in 1 hour. *(3 marks)*
  b) What common name is given to this quantity of energy? *(1 mark)*

7 Cartridge fuses are normally available in 3 A, 5 A or 13 A.
  a) What could happen if you used a 3 A fuse in the plug for a 3 kW electric fire? *(1 mark)*
  b) Why is it bad practice to use a 13 A fuse in a plug for a 60 W study lamp? *(1 mark)*
  c) What size of fuse would you use for a hairdryer labelled 230 V, 800 W? Explain how you worked out your answer. *(2 marks)*

F **8** What is the highest number of 60W bulbs that can be run off the 230V mains if you are not going to overload a 5A fuse? *(3 marks)*

9 A 13A socket is designed to allow a current of 13A to be drawn safely from it. Mr White connected the following appliances to a single 13A socket using a 4-way extension lead: 2.4kW electric kettle; a 3kW dishwasher; an 800W television; a 1300W toaster.
 a) Calculate the current through each appliance, assuming that the supply voltage is 230V. *(4 marks)*
 b) Assuming that the plug from the extension lead contained a 13A fuse, what would happen if he attempted to use all the appliances at the same time? *(1 mark)*

10 a) Copy the circuit diagram below and complete the two-way switches A and B so that the bulb lights. *(2 marks)* **F**

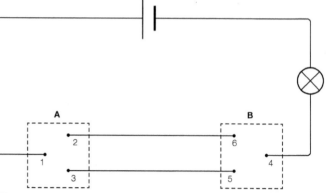

Figure 11.12

 b) Now that you have completed the switches, describe the path of the current from the positive terminal of the battery to the negative terminal by copying and completing the following sentence.
 The current flows from the positive terminal to 1 to ___ to ___ to ___ to the lamp and to the negative terminal. *(2 marks)*

Magnetic field pattern around a bar magnet

Magnetic fields can be investigated using a small **plotting compass**. The 'needle' in the compass is a tiny magnet which is free to turn on its spindle. When held near a bar magnet, the needle is turned by forces between its poles and the **poles of the magnet**. The needle comes to rest so that the turning effect is zero. Figure 12.1 shows how a plotting compass can be used to plot the field around a bar magnet.

Starting with the compass near one end of the magnet, the needle position is marked using two dots. The compass is then moved so that the needle lines up with the previous dot. Another dot is added and so on. When the dots are joined up, the result is a **magnetic field line**. More lines can be drawn by starting with the compass in different positions. In Figure 12.2, a **magnetic flux** (a selection of field lines) have been used to show the magnetic field around a bar magnet. It should be noted that:

▶ The field lines run from the North pole (N) to the South pole (S) of the magnet.

▶ The field direction, shown by the arrowhead, is defined as the direction in which the force on a North pole would act. The N end of the compass needle would point in this direction.

▶ The magnetic field is strongest where the field lines are closest together at the poles of the magnet. There are no field lines in the middle of the magnet, so there is no magnetic field here.

If two magnets are placed near each other, their magnetic fields combine to produce a single field.

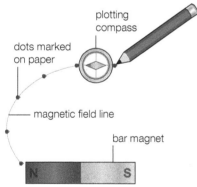

▲ **Figure 12.1** Plotting the field around a bar magnet using a compass

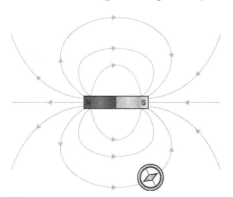

▲ **Figure 12.2** Magnetic field lines around a bar magnet

Magnetic field pattern due to a current-carrying coil

A coil with one turn

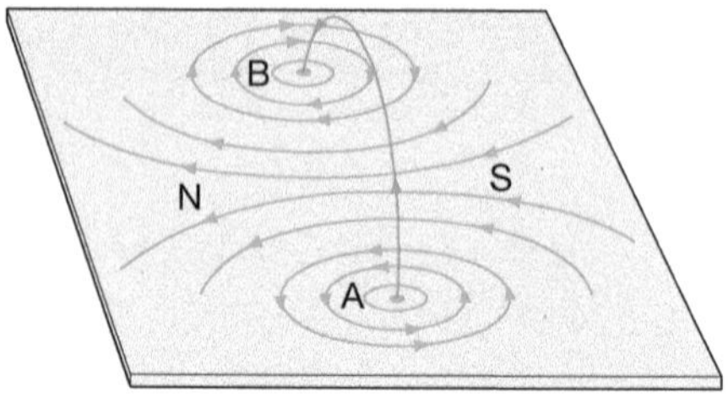

Figure 12.3 shows the magnetic field around a single loop of wire which is carrying a current. Near A, the field lines point anticlockwise and near B, the lines point clockwise. In the middle, the fields from each part of the loop combine to produce a magnetic field running from left to right. This loop of wire is like a very short bar magnet. Magnetic field lines come out of the left-hand side (North pole) and go back into the right-hand side (South pole).

The strength of the magnetic field can be increased by:

▶ increasing the number of turns of wire in the coil
▶ increasing the current through the coil.

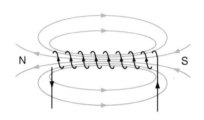

▲ **Figure 12.3** The magnetic field around a single loop of current-carrying wire

A coil with many turns

A stronger magnetic field can be made by wrapping a wire into a long coil. This coil is referred to as a **solenoid**. A current must be passed through it, as illustrated in Figure 12.4.

The magnetic field produced by a solenoid has the following features:

▶ The shape of the field is similar to that around a bar magnet and there are magnetic poles at the ends of the coils.
▶ Increasing the current increases the strength of the field.
▶ Increasing the number of turns on the coil increases the strength of the field.

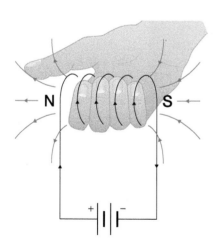

▲ **Figure 12.4** The pattern of magnetic field around a solenoid

Polarity

The **right-hand grip rule**, as shown in Figure 12.5, can be used to help you work out which way round the poles are in a solenoid. Imagine gripping the solenoid with your right hand, so that your fingers point in the direction of the conventional current. Your thumb will point towards the North pole of the solenoid.

▲ **Figure 12.5** The right-hand grip rulet

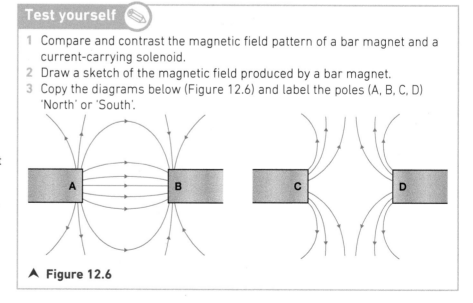

Test yourself 🖊

1 Compare and contrast the magnetic field pattern of a bar magnet and a current-carrying solenoid.
2 Draw a sketch of the magnetic field produced by a bar magnet.
3 Copy the diagrams below (Figure 12.6) and label the poles (A, B, C, D) 'North' or 'South'.

▲ **Figure 12.6**

Show you can ?

a) Plot the field pattern of a single bar magnet using the technique described on page 165. Then, using two bar magnets in the positions shown in Figure 12.7, verify that the field patterns would be as illustrated in the diagram.

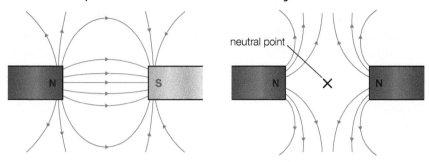

▲ **Figure 12.7**

b) Explain the meaning of the term 'neutral point'.

Prescribed practical

Investigating the factors affecting the strength of an electromagnet

An electromagnet is a solenoid wrapped around a soft iron core. The strength of an electromagnet, in other words the strength of its magnetic field, is measured by finding the mass of iron the electromagnet will attract. Iron nails or paper clips may be used.

Three factors can vary the strength of the electromagnet and these are discussed in turn below.

1 Investigating the effect of the current on the strength of the magnetic field

Apparatus
- a thick insulated coil of copper wire
- a soft iron core
- an ammeter
- a supply of iron nails
- a variable power supply

Method
1 Construct an electromagnet using 50 turns of insulated wire around a soft iron core.
2 Connect it to the circuit as shown in Figure 12.8.
3 Using the rheostat, increase the current in steps, measuring the number of iron nails attracted to the electromagnet for each current.
4 Record your results in a suitable table, similar to Table 12.1 on the next page.
5 Plot a graph of number of nails lifted on the *y*-axis versus current on the *x*-axis.
6 What does this graph show?

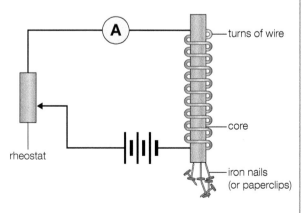

▲ **Figure 12.8** Investigating the strength of an electromagnet

Table 12.1 Results table

Current/A	0.0	0.5	1.0	1.5	2.0	2.5	3.0	3.5
Number of nails lifted								

A graph similar to Figure 12.9 should be obtained.

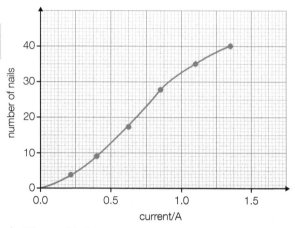

▲ **Figure 12.9** A typical graph of results

2 Investigating the effect of the number of turns on the strength of the magnetic field

Apparatus
- a thick insulated coil of copper wire
- a soft iron core
- an ammeter
- a supply of iron nails
- a variable power supply

Method
1 Keeping the current at 2.5 A and the material of the core constant, increase the number of turns of wire in steps of 10.
2 For each number of turns, measure the number of nails lifted.
3 Record your results in a suitable table, similar to Table 12.2.
4 Plot a graph of number of turns versus number of nails lifted.
5 What does your graph tell you?

Table 12.2 Results table

Number of turns	10	20	30	40	50	60
Number of nails lifted						

3 Investigating the effect of the material of the core on the strength of the magnetic field

Apparatus
- a thick insulated coil of copper wire
- cores made from various materials (at least: soft iron, steel, copper, plastic and wood)
- a supply of iron nails
- a power supply

Method
1 For this third investigation, keep the number of turns of wire and the current constant at 3.0 A but change the material of the core from soft iron to steel, copper, plastic, wood or air (no material).
2 In each case, measure the number of iron nails lifted and record your results in a table.
3 Draw a bar chart of your results and describe your findings.

Table 12.3 Results table

Type of material	Soft iron	Steel	Copper	Plastic	Wood	No material (air)
Number of nails lifted						

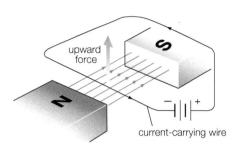

upward
force

S

N

current-carrying wire

▲ **Figure 12.10** The movement of a current-carrying wire in a magnetic field

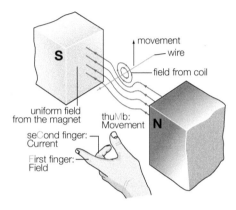

movement
wire
field from coil

S

N

uniform field from the magnet

thuMb: Movement

seCond finger: Current

First finger: Field

▲ **Figure 12.11** Fleming's left-hand rule

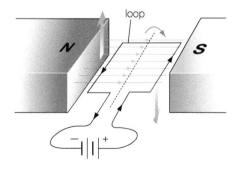

loop

N

S

▲ **Figure 12.12** When a current is passed through the loop, it experiences a turning effect

Magnetic force on a current-carrying wire: the motor effect

Figure 12.10 shows a length of copper wire that has been placed in a magnetic field. The copper wire is not magnetic so the wire itself is not affected by the magnet. However, with a current passing through it, there is a force on the wire. Where does this force come from?

Fleming's left-hand rule

The force arises because the current produces its own magnetic field which interacts with the field of the magnet. The resulting magnetic field is shown in Figure 12.11. Note how the field lines that were originally circular have become distorted due to the current in the wire. The field lines below the wire are concentrated, while the field lines above the wire are not. The result is that the wire experiences a force as the field lines tend to straighten.

If the current in the wire changes direction, or the polarity of the magnet is reversed, then the direction of the force on the wire is reversed.

The force is increased if:

▶ the current in the wire is increased

▶ a stronger magnet is used

▶ the length of the wire exposed to the magnetic field is increased.

The relationship between the direction of motion, the current and the magnetic field when a current-carrying wire is in a magnetic field is predicted by **Fleming's left-hand rule**. The rule states that if the thumb and first two fingers of the left hand are held at right angles to each other, then the thumb points in the direction of the force or motion, the first finger will be pointing in the direction of the field and the second finger will be pointing in the direction of the current.

When applying this rule, it is important to remember how the field and current directions are defined:

▶ The field direction is from the N pole of a magnet to the S pole.

▶ The conventional current direction is from the positive (+) terminal of a battery round to the negative (−).

Fleming's left-hand rule only applies if the current and field directions are at right angles to each other. If the current and magnetic field are parallel, there is no force on the wire and it will not move.

Several devices use the fact that there is a force on a current-carrying conductor in a magnetic field, such as loudspeakers and electric motors.

The turning effect on a current-carrying coil in a magnetic field

The loop in Figure 12.12 lies between the poles of a magnet. The current flows in opposite directions along the two sides of the loop. If you apply Fleming's left-hand rule, when a current is passed through the loop, one side of the loop is pushed up and the other side is pushed down. In other words, there is a turning effect on the loop.

If the number of loops is increased to form a coil, the turning effect is greatly increased. This is the principle involved in **electric motors**.

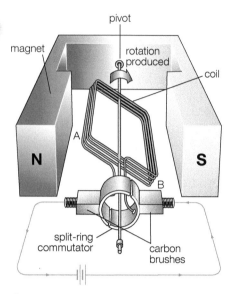

▲ **Figure 12.13** A simple electric motor

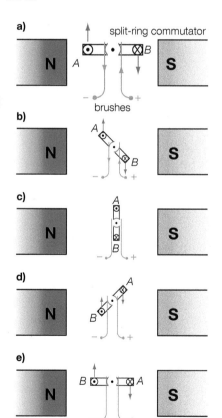

▲ **Figure 12.14** The turning action of the coil in the electric motor

The d.c. electric motor

Figure 12.13 shows a simple electric motor. It runs on **direct current (d.c.)**, the current that flows in one direction only from a battery. The coil is made of insulated copper wire. The coil is free to rotate on an axle between the poles of a magnet. The **commutator**, or split-ring, is fixed to the coil and rotates with it. Figure 12.14 details this action. The brushes are two contacts that rub against the commutator and keep the coil connected to the battery. They are usually made of carbon.

When the coil is horizontal, the forces are furthest apart and have their maximum turning effect on the coil. With no change to the forces, the coil would eventually come to rest in the vertical position. However, as the coil overshoots the vertical, the commutator changes the direction of the current through it. So the forces change direction and push the coil further around until it is again vertical, and so on. In this way the coil keeps rotating clockwise, half a turn at a time. If either the battery or the poles of the magnet are reversed, the coil will rotate anticlockwise.

The turning effect on the coil can be increased by:

▶ increasing the current in the coil
▶ increasing the number of turns on the coil
▶ increasing the strength of the magnetic field
▶ increasing the area of the coil.

Alternating and direct currents

A direct current (d.c.) always flows in the same direction: from a fixed positive terminal to the fixed negative terminal of a supply.

A typical d.c. circuit is shown in Figure 12.15. A cell or battery gives a constant, steady, direct current. A graph of voltage versus time for a d.c. supply is shown in Figure 12.16. The current is described as being unidirectional.

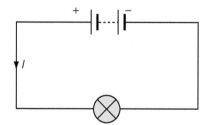

▲ **Figure 12.15** A simple d.c. circuit

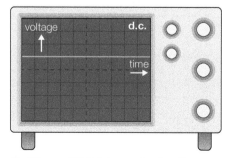

▲ **Figure 12.16** A graph of voltage against time for a d.c. supply

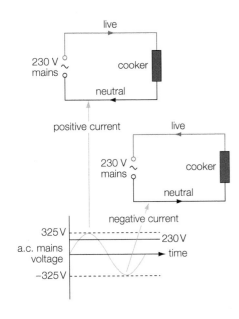

Figure 12.17 An a.c. supply

The electricity supply in your home is an **alternating current (a.c.)** supply. In an a.c. supply, the voltage changes size and direction in a regular and repetitive way (Figure 12.17).

In fact, the mains voltage changes from +325V to −325V. Considering the sizes of positive and negative voltages together, the effective value of the voltage is 230V. The current changes direction 100 times every second and makes 50 complete cycles per second; hence the frequency of the mains is 50Hz.

It is clear from Figure 12.18 why an a.c. supply is said to be bidirectional.

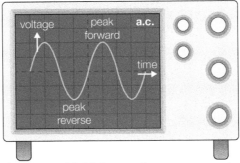

Figure 12.18 A trace for an a.c. supply

Test yourself

4 Using a diagram, state Fleming's left-hand rule.
5 Use Fleming's left-hand rule to complete the following exercises.
  a) Copy the diagrams in Figure 12.19. Draw arrows on parts i) and ii) to show the directions of the forces on the current-carrying wires.

⊗ Means current is perpendicular to and into the plane of the paper
⊙ Means current is perpendicular to but out of the plane of the paper

Figure 12.19

  b) Copy the diagrams in Figure 12.20 and indicate the direction of the current in the wire and the direction of the magnetic field in part i). In part ii), indicate the direction of the magnetic field and label the polarity of the magnets.

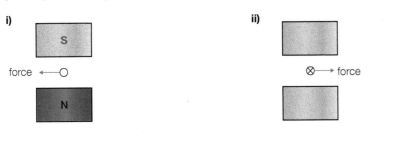

Figure 12.20

F

6 Look at Figure 12.21, which shows five displays on a cathode ray oscilloscope (CRO) screen. Which of the displays illustrate:
a) d.c. voltages
b) a.c. voltages?

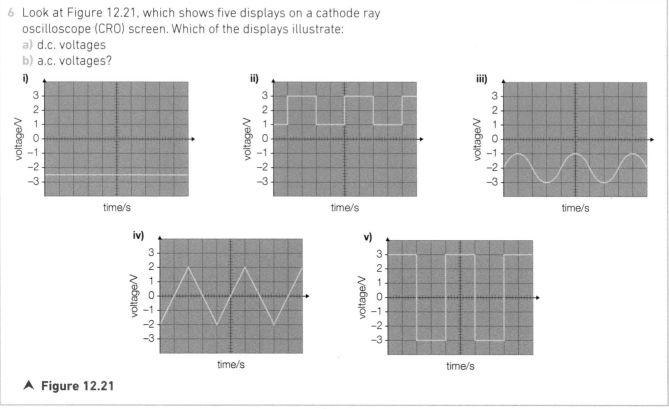

▲ **Figure 12.21**

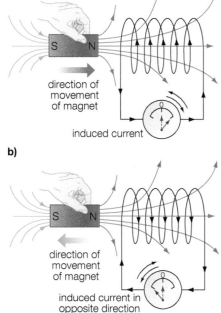

direction of movement of magnet

induced current

direction of movement of magnet

induced current in opposite direction

▲ **Figure 12.22** Inducing a current by moving the magnetic field

Electromagnetic induction

Electromagnetic induction (EMI) is the reverse of the motor effect. Michael Faraday is generally credited with its discovery in 1831. In EMI, the user supplies a magnetic field and motion to produce a voltage that can cause a current.

The easiest way to demonstrate electromagnetic induction is with a magnet and a coil of wire. A centre-zero ammeter can then be used to measure the induced current and detect the direction in which it is flowing. In Figure 12.22, a current is generated when the magnet is moving into or out of the solenoid. No current is generated when the magnet and coil are stationary.

We can predict the direction of the current flowing in the coil using **Lenz's Law**: the current always flows in the direction that will set up a magnetic field to oppose the motion causing it. In Figure 12.22a, the current flowing in the coil produces a North pole at the left-hand end of the coil. As the magnet moves towards the solenoid, there is a magnetic force that repels it, so you have to do some work to push the magnet into the solenoid. This pushing generates the electrical energy.

In Figure 12.22b, the magnet is being pulled out of the solenoid. The direction of the current is reversed and so there is now an attractive force acting on the magnet. The hand pulling the magnet is still doing work to produce electrical energy.

It is possible, although much less convenient, to induce a current without moving the magnet, but by moving the coil instead. So, in Figure 12.22a we would get the same effect as moving the magnet to the right by moving the coil to the left. The important idea is that there must be relative movement between the magnet and the coil.

When a current is produced by electromagnetic induction, energy is always needed to create the electrical energy. In the example described in Figure 12.22, the energy originally came from the muscles working the hand holding the magnet. So, electromagnetic induction is a way of converting mechanical energy to electrical energy. This is the principle behind transformers and electricity generation in our power stations. We will come back to this idea later in this chapter, in the section on the a.c. generator.

What happens if the coil is not part of a closed loop? Then a moving magnet induces a voltage across the ends of the coil, but because there is no complete circuit, no current flows.

Faraday made three important discoveries about EMI. He found out that he could increase the size of the induced current by:

▶ moving the magnet faster
▶ using a stronger magnet
▶ using a coil with a greater number of turns.

All three of the above have the effect of increasing the rate at which the magnetic field linked with the coil is changing.

It is also possible to induce an electric current by rotating a magnet within a coil of wire. This is the principle of the bicycle dynamo (Figure 12.23). The dynamo spindle turns when the wheels of the bicycle are rotating. The output from this type of dynamo is alternating current.

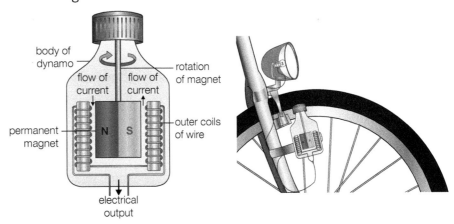

▲ **Figure 12.23** How a current is induced in a bicycle dynamo

EMI can also be demonstrated by moving a single wire between the opposite poles of a horseshoe magnet.

By carrying out the investigation shown in Figure 12.24, it can be shown that generation of current depends on the direction of movement of the wire. To generate a current, the wire must cross the magnetic field lines. A current is produced if the wire is moved up and down along the direction XX', but there is no current if the wire moves along ZZ' or YY'. Reversing the direction of movement reverses the current.

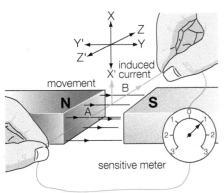

▲ **Figure 12.24** Electromagnetic induction using a single wire

So, if moving the wire up makes the pointer in the meter move to the right, then moving the wire down will make it go to the left.

The size of the current generated can be increased in the following ways:

▶ moving the wire more quickly
▶ using a stronger magnet
▶ looping the wire so that several turns of wire pass through the poles.

It is also possible to induce a current with this apparatus by keeping the wire stationary and moving the magnet. In this case, the magnet would have to move parallel to the line XX'.

Test yourself

7 Describe how you could demonstrate electromagnetic induction using a coil, a magnet and a centre-zero ammeter.
8 Describe the relationship between the polarity of the magnet and the direction of the induced current in a coil.

Show you can

a) Explain in terms of energy why there is no induced current if both the coil and the magnet are stationary.
b) Explain how you could produce alternating current by rotating a magnet.

Faraday's iron ring experiment

Michael Faraday's iron ring experiment is one of the classic experiments of the nineteenth century. Look at Figure 12.25. The soft iron ring in the centre has two coils wrapped around it. The coil on the right is connected to a switch and a small battery. The coil on the left is connected to a centre-zero ammeter. The coils are made of insulated copper wire, so no current can ever flow in the iron ring. In the situation shown in the diagram, there is no current flowing in either coil.

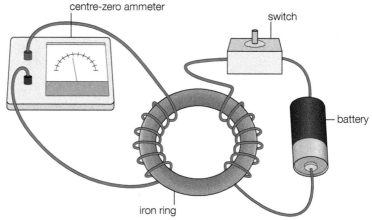

▲ **Figure 12.25** The apparatus needed for Faraday's iron ring experiment

What happens when the switch is closed? A current immediately flows in the coil on the right. But Faraday observed that there was a momentary deflection in the centre-zero ammeter. How could this be explained?

Faraday argued:

▶ Closing the switch caused an increase in the current in the right-hand coil.

▶ This caused an increasing magnetic field in the right-hand coil.

▶ Since the iron core links the two coils magnetically, there is an increasing magnetic field in the left-hand coil.

▶ The increasing magnetic field in the left-hand coil caused an induced current which flowed through the ammeter.

The effect is momentary; there is a deflection of the ammeter needle only when the current in the right-hand coil is changing. When there is a steady current in the right-hand coil, there is a steady magnetic field in both coils. For induction to occur we need the current in the left-hand coil to be changing!

What then happens when the switch is opened? There is a momentary deflection of the ammeter needle, but this time it is in the opposite direction, because the field in the left-hand coil is now decreasing.

This simple experiment is the basis for the transformer, without which electrical energy could not be transmitted efficiently from a power station to our homes.

Tip ↻

In 1824, Faraday was made a fellow of the Royal Society for his work on electricity. It is reported that the Prime Minister of the day, Mr Gladstone, asked Faraday what good would come of his discovery of electromagnetic induction. Faraday is said to have answered softly, 'Why Prime Minister, someday you can tax it.'

Test yourself ✎

9 a) Look at Figure 12.26. If the magnet is stationary above the coil, the reading on the centre-zero ammeter is zero. When the North pole of the magnet moves into the coil, the needle moves to the right. When the magnet is stationary inside the coil, the needle on the ammeter returns to zero.
   i) What name is given to this effect?
   ii) Explain these observations.
  b) Describe what would be observed when the magnet is pulled up out of the coil.
  c) What energy provides the source for the electrical energy in the coil?

10 Figure 12.27 shows the apparatus used in Faraday's iron ring experiment. The coils wrapped around the iron ring are made of insulated copper wire.
  a) Explain why a current flows in the secondary coil when the switch is closed.
  b) Explain why a current flows in the secondary coil when the switch is then opened.
  c) In what way is the current in the secondary coil different in parts a) and b)?
  d) How could the apparatus be changed to produce an a.c. current in the secondary coil?

11 State the relationship between the size of the induced current and the speed of the magnet.

12 State the relationship between the direction of the induced current and the direction of the magnet's motion.

13 State how alternating current can be induced using a coil and a magnet.

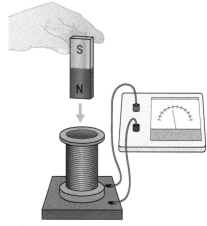

▲ **Figure 12.26**

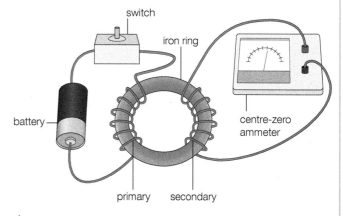

▲ **Figure 12.27**

The a.c. generator

Figure 12.28a shows the design of a simple **a.c. generator**. It only produces alternating current, so it is called an **alternator**.

Turning the axle makes the coil rotate in the magnetic field. The coil is shown with a single turn to keep the diagram as simple as possible. The rotation of the coil causes a voltage to be induced across the ends of the coil.

As the coil turns, the slip rings rotate. In contact with the slip rings are stationary brushes (marked 1 and 2 in the diagram). These provide a continuous contact between the coil and the cathode ray oscilloscope (CRO).

Figure 12.28b shows how the induced voltage changes with time. This can be observed on the CRO. Figure 12.28c shows the position of the coil at various times.

i) The coil is vertical. Sides AB and CD are moving parallel to the field lines, so no voltage is induced.

ii) The coil is horizontal. In this position, sides AB and CD are cutting the field lines at the greatest rate, so the induced voltage is a maximum.

iii) The coil is vertical once more, so there is no induced voltage.

iv) The coil is horizontal once more, but notice that AB and CD are moving through the field in the opposite direction when compared with position (ii). So, the induced voltage is once again a maximum, but now it is in the opposite direction.

v) The coil is vertical once more, so there is no induced voltage.

The size of the induced voltage can be made larger by:
▶ rotating the coil faster
▶ using stronger magnets
▶ using a coil of more turns
▶ wrapping the coil around a soft iron core.

a)

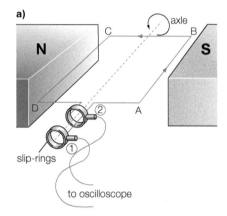

b)

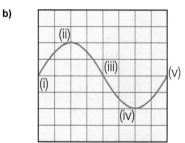

c)

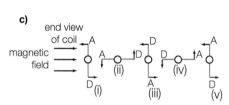

▲ **Figure 12.28** a) An a.c. generator b) How the voltage waveform produced by the generator appears on an oscilloscope screen c) The position of the coil

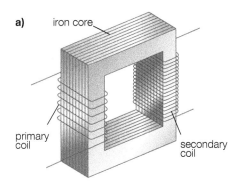

a) iron core

primary coil

secondary coil

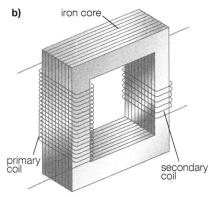

b) iron core

primary coil

secondary coil

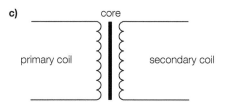

c) core

primary coil

secondary coil

▲ **Figure 12.29 a)** A step-up transformer **b)** A step-down transformer **c)** The circuit symbol for a transformer

Transformers

A transformer works only with alternating current. Both the input and the output voltages are a.c.

Figure 12.29a shows the construction of a transformer. All transformers consist of two coils of wire wrapped round a laminated iron core. This is an example of a **step-up transformer**; it has fewer turns in the primary coil than in the secondary coil. A step-up transformer is used to increase the voltage.

Figure 12.29b shows the construction of a **step-down transformer**, one that has fewer turns in the secondary coil than in the primary coil. A step-down transformer is used to decrease the voltage.

Figure 12.29c shows the circuit symbol for a transformer.

All transformers have three parts:

▶ a **primary coil**; the incoming voltage V_p (voltage across primary coil) is connected across this coil

▶ a **secondary coil**; this provides the output voltage V_s (voltage across the secondary coil) to the external circuit

▶ a **laminated iron core**; this links the two coils magnetically.

There is no electrical connection between the two coils, which are constructed using insulated wire.

To step up the voltage by a factor of 10, there must be 10 times as many turns on the secondary coil as on the primary. The turns ratio tells us the factor by which the voltage will be changed.

Tip

According to the principle of conservation of energy, if the voltage is stepped up then the current must be stepped down and vice versa. The energy per second going into the transformer must equal the energy per second leaving the transformer.

There is an important equation, known as the **transformer equation**, which relates the two voltages V_p and V_s to the number of turns on each coil, N_p and N_s.

$$\frac{\text{number of turns in secondary coil}}{\text{number of turns in primary coil}} = \frac{\text{voltage across secondary coil}}{\text{voltage across primary coil}}$$

$$\frac{N_s}{N_p} = \frac{V_s}{V_p}$$

Example

A transformer is needed to step down the mains voltage at 240 V to supply 20 V. If the primary coil has 4800 turns, how many turns must the secondary coil have?

Answer

$V_p = 240\,V$; $V_s = 20\,V$; $N_p = 4800$;

$N_s = ?$

$N_s = \dfrac{20 \times 4800}{240}$

$\quad = 400$ turns

Power in transformers

Transformers are very efficient devices with an efficiency that is typically above 99%. At GCSE level, we can treat transformers as being 100% efficient. For such a transformer we can write:

input electrical power to primary coil = output electrical power from secondary coil

$$V_p I_p = V_s I_s$$

where

V_p = the voltage across the primary coil

I_p = the current in the primary coil

V_s = the voltage across the secondary coil

I_s = the current in the secondary coil

Electricity transmission

Electrical power is distributed around the country from power stations through a grid of high-voltage power lines. This is called the electricity distribution network. The electricity in overhead power lines is transmitted to our homes and industry at 275 kV or 400 kV. The high voltages used in electricity transmission are extremely dangerous, which is why the cables that carry the power are supported on tall pylons high above people, traffic and buildings. High voltages are used because this means that there is a low current in the cables and so less energy is wasted.

When a current flows in a wire or cable, some of the energy it is carrying is lost because of the cable's resistance; the cables get hot. A small current wastes less energy than a high current.

Consider the following two methods of transmitting electricity:

Method 1	Method 2
Let the resistance of the cable be $10\,\Omega$, while carrying a current of 1000 A. Then	Let the resistance of the cable be $10\,\Omega$, while carrying a current of 100 A. Then
$R = 10\,\Omega$	$R = 10\,\Omega$
$I = 1000\,A$	$I = 100\,A$
Power lost $= I^2 R = 1000^2 \times 10$	Power lost $= I^2 R = 100^2 \times 10$
$= 10\,000\,000\,W$	$= 100\,000\,W$
$= 10\,MW$	$= 0.1\,MW$

Clearly, transmitting the smaller current reduces the heat lost in the cables dramatically.

Before electricity reaches homes, the high voltage must be stepped down to a much safer 230 V. Today, almost all domestic electrical appliances in Europe operate at 230 V.

14 An electric current flows in a conductor in a magnetic field, as is shown in Figure 12.30. Copy the diagram and mark on it the direction of the force on the wire.

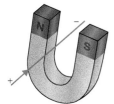

▲ **Figure 12.30** Current in a magnetic field

15 A bar magnet is moving towards a loop of wire as shown in Figure 12.31.
   a) Is a voltage induced in the wire? Give a reason for your answer.
   b) Is a current induced in the wire? Give a reason for your answer.
   c) The open ends of the wire are now connected to a torch bulb. Copy Figure 12.31, add the bulb and mark on your diagram the direction in which the current flows.

16 A coil of insulated copper wire wrapped around a ring of soft iron is connected to a battery via a switch (Figure 12.32). A coil on the other side of the ring is connected to a centre-zero ammeter. Describe and explain what is seen on the ammeter when:
   a) the switch is closed
   b) the switch is re-opened.

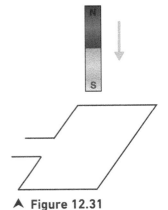

▲ **Figure 12.31**

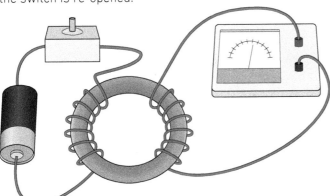

▲ **Figure 12.32**

17 A coil is connected to a cathode ray oscilloscope (Figure 12.33). When a bar magnet falls through the coil, two spikes are observed on the CRO. Describe and explain the appearance of the trace on the screen of the CRO.

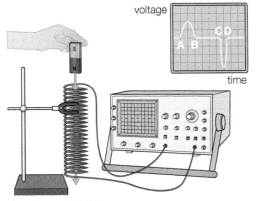

▲ **Figure 12.33**

18 An anemometer is a device to measure wind speed. Figure 12.34 shows a simple anemometer. When the wind blows, the plastic cups turn.
a) Explain why the wind causes the voltmeter to give a reading.
b) Explain why the reading on the voltmeter is a 'measure' of the wind speed.
The gauge is not sensitive enough to measure the speed of a gentle breeze.
c) Suggest one way that the design can be modified to make the anemometer more sensitive.

19 Jo makes a model generator using the apparatus shown in Figure 12.35. Jo pulls down the bar magnet and then lets go. The magnet oscillates into and out of the coil.
a) Explain why a current flows in the coil.
b) Is the current in the coil alternating or direct? Give a reason for your answer.
c) Copy the axes below (Figure 12.36) and sketch the graph of current against time for the model generator.

▲ **Figure 12.36**

20 Figure 12.37 shows how electricity from a power station is distributed to homes.

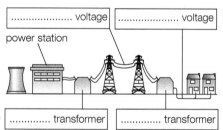

▲ **Figure 12.37** How electricity from a power station is distributed to homes

a) Copy the diagram and write the names of the types of transformer in the boxes.
b) In the appropriate boxes on your diagram, label those parts of the distribution system where the voltage is high and those parts where it is low.
c) Why are high voltages used in the national distribution of electricity?

21 At a power station, the main transformer is supplied from a 25 kV generator.
a) How much energy is transferred from the generator for each coulomb of charge?
b) The main transformer steps up the voltage to 275 kV before sending it out to the grid. Describe fully the purpose of stepping up the voltage.
c) In what other part of the electricity transmission system must transformers be used?
d) Why must these other transformers be used?

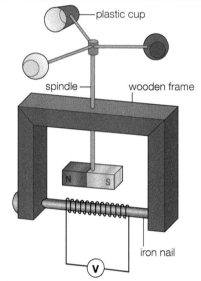

▲ **Figure 12.34** An anemometer

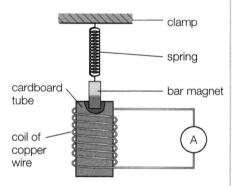

▲ **Figure 12.35** A model generator

22 Two coils are placed side by side. A current meter is connected to one coil as shown in Figure 12.38.

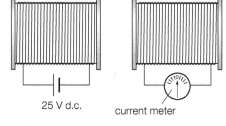

a) The first coil is connected to a 25 V d.c. supply. Describe what happens to the meter reading:
  i) when the first coil moves away from the second coil
  ii) when the first coil moves back towards the second coil.

b) The 25 V d.c. supply is now replaced with a 25 V a.c. 1 Hz supply connected to the first coil. Describe what (if anything) happens to the meter reading if the first coil remains stationary.

▲ **Figure 12.38** Two coils

23 Figure 12.39 shows a transformer. The input electrical power to the primary coil is 7.2 kW.

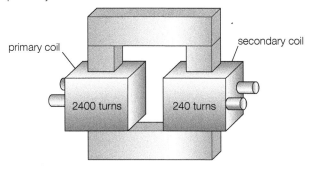

▲ **Figure 12.39** A transformer

a) Calculate the voltage at the secondary coil if the voltage across the primary coil is 8 kV.

b) Assuming the transformer has an efficiency of 100%, calculate the maximum current which can be drawn from the secondary coil.

24 State the function of:
a) a step-up transformer
b) a step-down transformer.

25 Explain why transformers are used at the power station and at the user substation in the transmission and distribution of electrical energy.

26 State the turns-ratio equation for transformers.

Show you can ?

a) Describe the design of a simple a.c. generator.
b) Draw the essential components of a simple transformer.

1 An electromagnet is a coil of wire through which a current can be passed.
   a) State three ways in which the strength of the electromagnet may be increased. *(3 marks)*
   b) An electromagnet may be switched on and off. Suggest one situation where this would be an advantage over the constant field of a permanent magnet. *(1 mark)*
   c) A coil carrying a current has two magnetic poles.

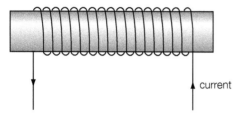

current

Figure 12.40

   i) Copy the diagram and mark the magnetic poles produced. *(2 marks)*
   ii) On your diagram, draw the magnetic field produced. *(4 marks)*

2 The motor effect may be demonstrated using the apparatus shown in the diagram below. When a current is passed through the moveable brass rod, it rolls along the fixed brass rails.

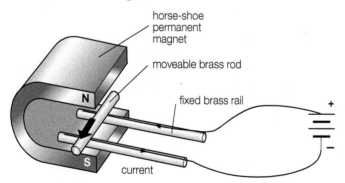

horse-shoe permanent magnet

moveable brass rod

fixed brass rail

N

S

current

Figure 12.41

   a) State the direction of the magnetic lines of force between the poles of the magnet. *(1 mark)*
   b) In which direction will the rod roll? *(1 mark)*
   c) The rod is placed back in its original position. What will happen to the rod if the poles of the magnet are reversed? *(1 mark)*
   d) The rod is placed back in its original position. What will happen to the rod if the poles of the battery are reversed? *(1 mark)*

3 Figure 12.42 shows a simplified diagram of a d.c. motor. The loop of wire is horizontal in a horizontal magnetic field.

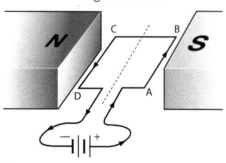

N   C   B   S

D   A

− +

Figure 12.42

   a) What does 'd.c.' mean? *(1 mark)*
   b) In which direction is the force on side AB of the wire loop? *(1 mark)*
   c) In which direction is the force on side CD of the wire loop? *(1 mark)*
   d) Explain how these forces cause the loop to rotate. *(1 mark)*
   e) What can you say about the force on side BC of the loop? *(2 marks)*

4 The diagrams show a bicycle dynamo.

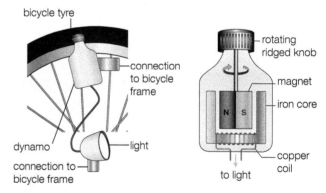

bicycle tyre

connection to bicycle frame

rotating ridged knob

magnet

iron core

N S

dynamo

light

connection to bicycle frame

copper coil

to light

Figure 12.43

   a) Explain fully why the lamp lights when the bicycle wheel turns. *(3 marks)*
   b) Why does the lamp get brighter as the bicycle moves faster? *(3 marks)*
   c) Why does the lamp not work at all when the bicycle stops? *(1 mark)*

5 The diagram shows a coil that can be rotated between the poles of a permanent magnet. The electrical output is at the connections to the brushes. The handle is turned vigorously.

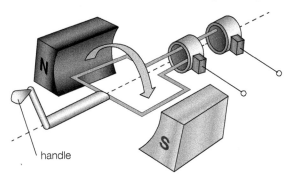

Figure 12.44

a) Is the electrical output a.c. or d.c.? *(1 mark)*
b) Explain why there is no electrical output when the plane of the coil is vertical. *(2 marks)*
c) Explain why the electrical output is greatest when the plane of the coil is horizontal. *(2 marks)*
d) State three ways in which the electrical output from this device could be increased. *(3 marks)*

6 The diagram shows an iron ring on which two coils are wound.

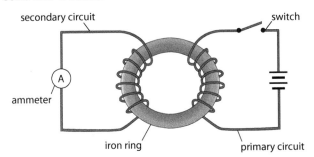

Figure 12.45

Describe and explain what would be observed on the ammeter when
a) the switch is closed *(2 marks)*
b) the closed switch is then opened. *(3 marks)*

7 Figure 12.46 shows a laboratory demonstration of how a transformer can be used to produce very large currents. Here the current is used to melt an iron nail.

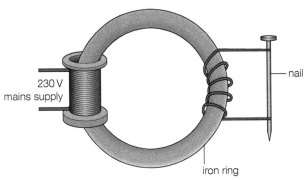

Figure 12.46

There are 500 turns in the coil on the left-hand side of the diagram and five turns in the coil on the right-hand side of the diagram.
a) Show that the voltage across the nail is 2.3 V. *(3 marks)*
b) If the current in the coil with 500 turns is 1.15 A, calculate the current in the nail, assuming the transformer has an efficiency of 100%. *(3 marks)*
c) Explain why the wires in the left-hand coil must be quite thin, but those in the right-hand coil must be very thick. *(2 marks)*

8 The diagram shows a simple transformer. The primary coil is connected to an a.c. supply.

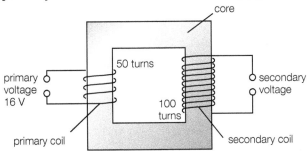

Figure 12.47

a) What do the letters a.c. stand for? How is a.c. different from d.c.? *(3 marks)*
b) The primary and secondary coils are both wound on the same core. What is this core made of? *(1 mark)*
c) Using the information shown on the diagram, calculate the secondary voltage. Show clearly how you get your answer. *(4 marks)*

13 Space physics

The Solar System

We now know that the Earth is one of eight **planets** that travel around the Sun. Each planet travels in an elliptical path and, with the exception of Mercury and Venus, they all have at least one moon.

Other objects also orbit the Sun. These are the comets and the asteroids. Table 13.1 gives some data about the planets.

Table 13.1 Some data on the eight planets orbiting the Sun

Planet	Planet diameter compared with Earth	Average distance of planet from Sun compared with Earth	Time to orbit the Sun compared with the Earth	Number of moons
Mercury	0.4	0.4	0.2	0
Venus	0.9	0.7	0.6	0
Earth	1.0	1.0	1.0	1
Mars	0.5	1.5	1.9	2
Jupiter	11.2	5.2	12.0	14
Saturn	9.4	9.5	29.0	24
Uranus	4.1	19.1	84.0	15
Neptune	3.9	30.1	165.0	3

Note that the status of Pluto was changed from a 'planet' to a 'dwarf planet' in 2005.

The orbits of the inner planets are almost circular, with the Sun at the centre. The orbits of Jupiter, Saturn, Uranus and Neptune are much more elliptical (like a rugby ball).

All the planets orbit the Sun in the same plane and travel in the same direction, as a result of the gravitational force between the Sun and the planets. This is evidence that they were formed at around the same time.

The four inner planets have a rocky surface on which it is possible to land a spacecraft. The four outer planets are known as gas giants. They are made up of very dense accumulations of gases, like hydrogen, methane and ammonia.

All of the planets except Mercury and Venus have at least one moon. **Moons** are heavenly bodies which are natural satellites of a planet. **Artificial satellites** are objects put into space by humans. Almost all orbit the Earth, but a few, like Kepler, orbit the Sun, and others, like the Mars Reconnaissance Orbiter, orbit other planets.

Artificial satellites of the Earth have four main purposes:

▶ communications
▶ Earth observation (spying and monitoring rainforests, deserts, crops etc.)
▶ astronomy
▶ weather monitoring.

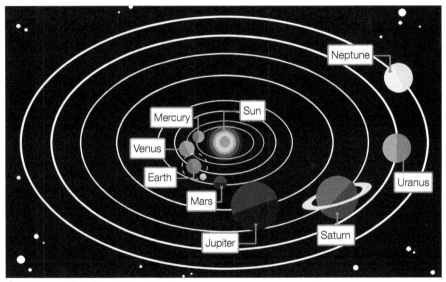

▲ **Figure 13.1** The Solar System

Comets

Comets range from a few hundred metres to tens of kilometres in diameter, and are sometimes called 'dirty snowballs'. At their centres are rock, ice, silicates and some organic compounds. Surrounding this is a 'coma' consisting of gases and dust.

Most comets orbit the Sun in very elliptical paths. As a comet approaches the Sun, solar radiation vaporises some of the frozen gas at the centre, causing the coma greatly to increase in size. Dust and gas stream away from the comet as a long tail, often millions of kilometres long. The radiation from the Sun causes this tail to point away from the Sun. Most astronomers believe that comets in our Solar System originate in the Oort cloud, a region of space between 2000 and 3000 times further from the Sun than Pluto.

Asteroids

Asteroids are large rocks in outer space. Some are very large, while others are as small as a few metres in diameter. Due to their small size, asteroids do not have enough gravity to pull themselves into the shape of a ball. Many are found in the Asteroid Belt, a giant ring between Mars and Jupiter.

Asteroids are left over materials from the formation of the Solar System. These materials were never incorporated into a planet because of the strong gravitational pull of Jupiter.

Test yourself 🖉

1 Name all the planets in order of their distance from the Sun, from closest to furthest.
2 Describe the major differences between the inner four planets and the other four.
3 Describe the differences between artificial and natural satellites, and give an example of each.
4 State three uses of artificial satellites of the Earth.
5 State the main features of a comet.

The life cycle of stars

Stars are formed in the cold clouds of hydrogen gas and dust known as stellar nebulae. Gravity causes these gas particles to come together. The gravitational force is greater than the outward pressure due to the particles' kinetic energy, and it brings about the gravitational collapse of the cloud. During this collapse, the material at the centre of the cloud heats up as the gravitational potential energy changes into thermal energy. The hot core at the centre of the cloud is called a protostar.

As the protostar accumulates more and more gas and dust, its density and temperature continue to rise, increasing the outward pressure within the protostar. A point is reached where this outward force is balanced by the gravitational force and the protostar becomes luminous because of its extremely high temperature (Figure 13.2).

If the mass of the protostar is greater than about 8% of our Sun's mass, the temperature will exceed the minimum required for nuclear fusion to begin. There is equilibrium between the inward gravitational force and the outward force from the radiation pressure due to fusion. The star is now in the main phase of its life, so it is called a main sequence star.

The smaller a star is, the longer its life as a main sequence star. Our Sun will have a life of around 10 billion years in the main sequence stage, while more massive stars only live for a few million years as main sequence stars. The life cycle of a star can be seen in Figure 13.3.

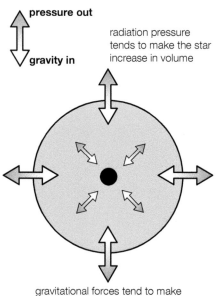

pressure out

gravity in

radiation pressure tends to make the star increase in volume

gravitational forces tend to make the star decrease in volume

▲ **Figure 13.2** Inward gravitational forces balance the force from radiation pressure in a star

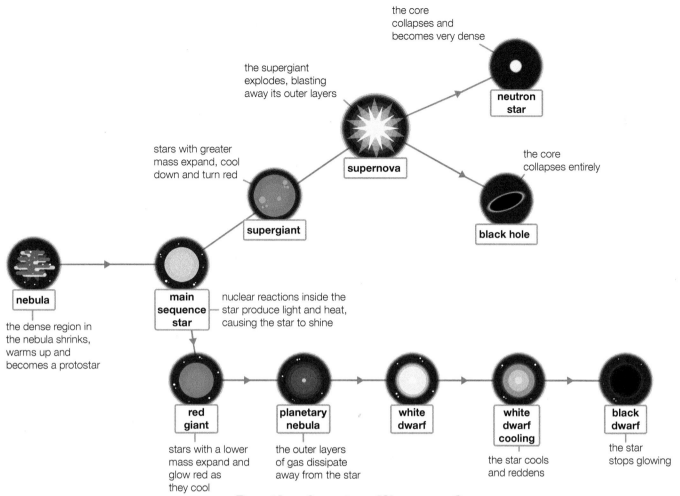

the core collapses and becomes very dense

neutron star

the supergiant explodes, blasting away its outer layers

stars with greater mass expand, cool down and turn red

supernova

supergiant

the core collapses entirely

black hole

nebula

the dense region in the nebula shrinks, warms up and becomes a protostar

main sequence star

nuclear reactions inside the star produce light and heat, causing the star to shine

red giant

stars with a lower mass expand and glow red as they cool

planetary nebula

the outer layers of gas dissipate away from the star

white dwarf

white dwarf cooling

the star cools and reddens

black dwarf

the star stops glowing

▲ **Figure 13.3** The life cycle of a star

Death of a star like our Sun

A stage is reached when almost all the hydrogen in the core of the star has been converted into helium by nuclear fusion. Without hydrogen, the energy output from fusion reactions in the core reduces so much that gravity compresses it significantly, but the star doesn't shrink.

Surrounding the core is a layer of hydrogen. Gravitational contraction provides enough energy for nuclear fusion of the hydrogen in this layer. The outward pressure from the nuclear fusion reactions prevents the star from collapsing, instead making it expand to several hundred times its former size. The surface temperature falls, and the starlight is now predominantly orange. We refer to the star as a '**red giant**'.

Within a red giant, other nuclear reactions can take place. Helium, for example, can fuse to become carbon and oxygen. Indeed, all the naturally occurring elements in the periodic table up to iron are formed by nuclear fusion in stars.

Close to the end of the life of a red giant, the gravitational force can no longer hold the outer layers of gas. These outer layers flow out, cool and surround the core to form a nebula. This nebula may eventually contribute to the creation of another star. Over time, the core that remains cools to become a **white dwarf**. Eventually, all fusion stops, and the star cools to become a **black dwarf**.

Show you can

a) State the main constituents of a stellar nebula.

b) Describe how a star forms from the material within a nebula.

c) Understand what is meant by nuclear fusion.

d) Explain why the size of our Sun has not changed in over 4 billion years.

e) Describe what will happen to our Sun when its hydrogen fuel is used up.

f) Explain what a red supergiant is.

g) Describe what a neutron star is.

h) Describe how black holes get their name, and explain why nothing can escape from them.

Death of a high mass star

If a star is more massive than the Sun, it goes through a slightly different process when the hydrogen fusion process ends. The rate at which helium fusion occurs is much more rapid than for a star of smaller mass. The huge amount of energy from helium fusion pushes the outer layers of the star outwards, and it turns into a **red supergiant**. Red supergiants are among the largest stars in the universe by volume, but not by mass.

Most red supergiants are several hundred times the radius of our Sun. They require huge amounts of energy to sustain them and to prevent gravitational collapse. As a result, they burn through their nuclear fuel very quickly, and most only live for a few tens of millions of years.

A red supergiant successively fuses different elements in the periodic table, up to the creation of iron. At that point, the core can no longer sustain outward radiation pressure, and the force of gravity causes the supergiant to begin to collapse.

This collapse releases gravitational potential energy that heats up the outer layers of the star. These are thrown off the star in an enormous explosion called a **supernova**. It is a really dramatic moment in the life of a massive star. For about a month, the supernova emits more radiation than all the other stars in its galaxy put together! For a relatively short time, the supernova shines with the brightness of ten billion Suns. The interaction of the elements exploding outwards from the supernova with atoms of elements and other particles surrounding the supernova is thought to produce the naturally occurring elements with atomic masses larger than iron. But the star's life is not quite over.

The core of the star is left behind, having been compressed into a **neutron star** by the immense gravitational pressure. For very massive stars, a **black hole** is created.

Neutron stars are composed almost entirely of neutrons, and are the smallest and densest stars known to exist. They typically have a radius of about 10 km, but they can have a mass of about twice that of the Sun.

Black holes are incredibly dense, and so have such enormous gravitational fields that nothing can escape from them, not even light. That is why we call them black holes. We can only infer their presence by the way in which they bend light passing close by.

Experimental evidence

What is the evidence that there is hydrogen, helium and so on in the Sun? The evidence comes from the absorption **spectrum** of sunlight (Figure 13.4). We are all familiar with the continuous spectrum of sunlight. A German physicist called Joseph von Fraunhofer (Figure 13.5) looked carefully at the spectrum and found that it contained many dark lines, which we now call 'Fraunhofer lines'.

What are the Fraunhofer lines and how are they formed?

When the visible light from below the Sun's surface passes through the layers above, the atoms in the solar 'atmosphere' absorb particular wavelengths, and so these wavelengths are missing in the spectrum we see. When there is no light at a particular wavelength, it appears as a black line in the spectrum (Figure 13.4).

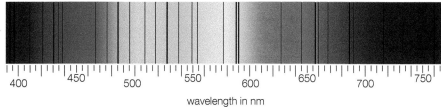

▲ **Figure 13.4** Solar absorption spectrum

Each of the black lines corresponds to an element in the Sun's atmosphere. The absorption spectrum tells us that at the moment, the major gases (by mass) in the Sun are hydrogen (71%), helium (27%), and oxygen (1%).

Nuclear fusion in our Sun

Stars are the powerhouses of the Universe. Stars like our Sun get their energy mainly from the fusion of light hydrogen nuclei into heavier helium nuclei. The electrical repulsion between the positive charges on the nuclei means that the nuclei must be moving very fast if fusion is to occur. In fact, fusion requires a temperature of at least 13 million °C and a density of $100\,g/cm^3$. Fortunately, these conditions are met at the core of our Sun.

The Doppler effect

Think about what we hear when a police car passes with its siren sounding. As the car approaches, the sound appears to have a higher pitch (or shorter wavelength) than we would expect. As soon as the car passes, its pitch falls. This is called the Doppler effect.

Look at Figure 13.6a, which shows a source of sound at rest. In Figure 13.6b, this source of sound is moving to the right. Observer A hears sound of high pitch (shorter wavelength) because the waves are being bunched up together. Observer B, to the left of the source, hears a sound of low pitch (longer wavelength) because the waves are being spread out.

A similar effect occurs with light. If the light that we observe from a moving source has a shorter wavelength than expected, it is because the source is moving towards us – we say the light is 'blue shifted'. But if the light we observe has a longer wavelength than expected, it is because the source is moving away from us – we say the light is 'red shifted'. Today, astrophysicists interpret the red shift from distant galaxies as evidence that space itself is expanding.

▲ **Figure 13.5** Joseph von Fraunhofer

a) source of sound at rest

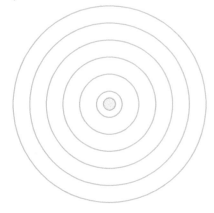

b) source of sound moving to the right

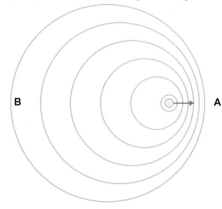

▲ **Figure 13.6** The Doppler effect

The origin of our Universe

Most physicists today believe that our universe began about 14 billion years ago with a 'Big Bang'. Evidence for the Big Bang theory comes from red shift.

Galaxies are huge collections of star systems. Our own galaxy, the Milky Way, contains over 100 billion stars! When we look at the light from distant galaxies, we find that it is shifted to the red (longer wavelength) end of the visible spectrum. This red shift is due to the Doppler effect.

How do we interpret these strange absorption spectrum data? If there is a red shift in the light from another galaxy, this tells us that the source is moving away from us. The fact that we almost always get red shift from the distant galaxies tells us that nearly all galaxies are moving away from us.

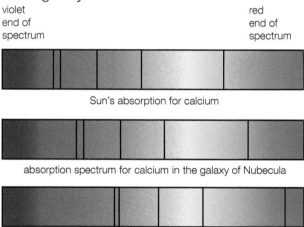

violet end of spectrum

red end of spectrum

Sun's absorption for calcium

absorption spectrum for calcium in the galaxy of Nubecula

absorption spectrum for calcium in the galaxy of Leo

▲ **Figure 13.7** Red shift in calcium from different galaxies

The red shift in the absorption spectrum for calcium in Figure 13.7, for example, tells us that the galaxies Nubecula and Leo are both moving away from us, and that Leo is moving away faster than Nubecula. If nearly all the galaxies are moving away from each other, we can infer that the universe (space itself) is expanding.

There is a further piece of evidence supporting this. In 1964, two American astrophysicists, Penzias and Wilson, detected microwaves of wavelength 7.35 cm that did not come from the **Earth,** the **Sun** or our closest stars. The microwaves were evenly spread over the sky, and were present day and night. They concluded that these microwaves were coming from outside our own galaxy. These waves come from the cosmos.

Today, we call that radiation cosmic microwave background radiation (CMBR). It represents the signature or 'afterglow' of the Big Bang that occurred 14 billion years ago. Currently, the only model that can give an explanation for CMBR is the Big Bang theory.

Tip

The words cosmic, background, microwave and radiation are all important when talking about CMBR. Be sure to give all four words in your answers!

The Big Bang theory

The argument begins by suggesting that the reason all the galaxies are currently moving away from each other is that they all originated from a common point, called a singularity. About 14 billion years ago, the Universe came into existence suddenly with an enormous explosion, which we call the Big Bang. It immediately went into a short period of very rapid growth, known as **inflation**.

As the Universe expanded, it cooled down and became less dense. Cooling allowed subatomic particles such as neutrons and protons and electrons to form. As the cooling continued, protons and neutrons combined to form simple nuclei. Eventually, around 380 000 years after the Big Bang, and after further expansion and cooling, the first stars came into existence.

Exoplanets

Exoplanet is the name given to planets outside our Solar System. Today, there are over 3600 known exoplanets in over 2600 known solar systems. There are two common ways by which astrophysicists have detected exoplanets – the radial velocity method, and the transit method.

Radial velocity method

We commonly think that a planet orbits a star, but this is not exactly so. Think of the star and the planet as a single system. Together, they have a common centre of gravity (strictly, a common centre of mass) somewhere between them. As the star is much more massive than the planet, the common centre of gravity will be much closer to the star than the planet. Both the star and the planet go around the common centre of mass. The star orbits as well as the planet! So relative to the planet, sometimes the star is moving towards us, and sometimes it is moving away from us.

This additional movement of the star around the common centre of mass with the exoplanet leads to small variations in the velocity of the star as observed from Earth. This in turn leads to tiny variations in the wavelengths of the light from that star due to the Doppler effect. So precise are the devices used today to observe stars that variations in velocity as small as 1 m/s can be detected and used to infer the existence of an exoplanet.

Transit method

Suppose an exoplanet in orbit around a star passes between us and the star. This is called a **transit**. There is a very small (typically about 1%) but detectable drop in the brightness of that star as seen from Earth. This is illustrated in Figure 13.8. In this way, transits are used to infer the existence of exoplanets.

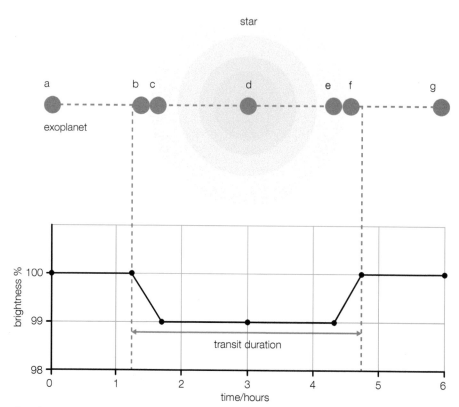

▲ **Figure 13.8** The transit of a large planet in front of a star

Planetary atmospheres

We said earlier that we could identify hydrogen and helium in the region above the surface of the Sun by looking at the absorption spectrum of the light passing through it. We went on to say that the black Fraunhofer lines were indicators of the elements contained within the solar 'atmosphere'.

A very similar technique is used to identify the elements in planetary atmospheres. The difference is that planets themselves, unlike the Sun, do not emit their own light. The planets do, however, scatter sunlight from their surfaces – and that sunlight then passes through the planetary atmosphere and out into space. When we examine that light, the black lines in the absorption spectra allow us to identify the elements in the atmosphere. Table 13.2 shows the atmospheres of the rocky planets.

Table 13.2 Atmospheres of the rocky planets

Planet	Atmosphere	
Mercury	Almost none at all	
Venus	Carbon dioxide	96%
	Nitrogen	3.5%
Earth	Nitrogen	77%
	Oxygen	21%
	Argon	1%
	Water vapour	variable
Mars	Carbon dioxide	95%
	Nitrogen	2.7%
	Argon	1.6%

Of particular importance to physicists studying planetary atmospheres is the existence of oxygen and water. Oxygen and water are essential for human life, and their presence in the atmosphere of another planet might indicate the possibility of life there.

The speed of light

Most people are aware that light is incredibly fast. But just how fast is it? Nothing can travel faster than light. Light travels at 300 000 km/s (3×10^5 km/s or 3×10^8 m/s). That might not sound very fast, but light can travel:

- more than 7 times around the Earth in just 1 second
- from the Earth to the Moon and back in less than 3 seconds
- from the Sun to Earth in about 500 seconds
- from Earth to Neptune in less than 5 days when the planets are furthest apart.

Light is very fast, but astronomical distances are also enormous. That is why astronomers sometimes measure distances in light years. A light year (ly) is the distance light travels in one year.

So, just how far is 1 ly?

distance = speed × time

$= 3 \times 10^8$ m/s $\times (60 \times 60 \times 24 \times 365)$ s

$= 9.46 \times 10^{15}$ m

$= 9.46 \times 10^{12}$ km

= over 63 000 times the distance from the Earth to the Sun

The nearest star system to Earth (other than our Sun) is Alpha Centauri. That is 4.2 ly away from Earth. So even if we could travel at the speed of light, it would take us 4.2 years to reach Alpha Centauri.

Most exoplanets are much further away than Alpha Centauri, so we are not going to reach them anytime soon. Our fastest spacecraft can travel at around 70 000 m/s. At this speed, it would take a staggering 18 000 years to reach Alpha Centauri or any exoplanet orbiting it.

The vast distances to the stars mean that with our present technology, it is not currently feasible to visit any planet outside our Solar System. There are many difficulties:

- flight time – the distance is so great that the flight would last for many generations
- engineering – our spacecraft are just too slow
- logistics – it is not clear how the spacecraft could carry enough fuel, oxygen and water
- ethics – the chance of failure would be high, with no possibility of return to Earth or rescue.

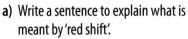

Show you can   (?)

a) Write a sentence to explain what is meant by 'red shift'.

b) Explain what red shift tells us about neighbouring galaxies.

c) Describe the experimental evidence that there is hydrogen in the Sun.

d) Explain the significance of CMBR.

e) Describe how the first atoms came into existence after the Big Bang.

f) Explain what is meant by nuclear fusion, and where it occurs naturally in the Universe.

g) Write down, according to current estimates, the maximum age of the visible Universe.

1 Figure 13.9 shows a cloud of gas and dust known as a nebula. The bright spots are stars.

Figure 13.9

   a) What force causes the gas to form stars? *(1 mark)*
   b) What two gases are the main constituents of stars? *(1 mark)*
   c) How do astronomers know this? *(1 mark)*
   d) Name the process that supplies the energy in stars. *(1 mark)*
   e) Apart from producing energy in stars, what else is produced by this process? *(1 mark)*

2 There are five stages in the lifecycle of a star with the same mass as our Sun.
   a) Copy and complete the list below. One stage has been inserted for you.
   _____, main sequence star, _____, _____, _____ *(2 marks)*
   b) Why do main sequence stars, such as our Sun, remain stable for many billions of years? *(2 marks)*
   c) What type of star results in a supernova? *(2 marks)*
   d) Some stars end their life as a black hole. Why is this stage of a star's life given this name? *(1 mark)*

3 The speed with which a galaxy is moving away from us is called its recession speed. The table shows how the recession speeds of different galaxies change with distance from the Earth.

Recession speed/m/s	450	700	920	1150	1380	1610
Distance/m	2×10^{20}	3×10^{20}	4×10^{20}	5×10^{20}	6×10^{20}	7×10^{20}

   a) On graph paper and starting from a (0,0) origin, plot a graph of recession speed (vertical axis) against distance (horizontal axis), and draw a straight line of best fit through the data points. *(5 marks)*

   b) Does your graph show that the recession speed is directly proportional to distance? Give a reason for your answer. *(2 marks)*
   c) The gradient of this graph is known as the Hubble constant, H_0. Show that the Hubble constant is approximately 2.3×10^{-18}/s. *(3 marks)*
   d) The Canis Major dwarf galaxy is approximately 2.4×10^{20} m away from Earth. Use the Hubble constant (or your graph) to show that this galaxy is moving away from us at approximately 2000 kilometers per hour. *(3 marks)*

4 Discuss briefly one observation which indicates that there are planets beyond our Solar System. *(3 marks)*

5 State the main difficulties that physicists and engineers would have in using manned spacecraft to explore any planet outside our Solar System. *(4 marks)*

6 Describe what happens to a very massive star once it passes out of the main sequence period of its lifecycle. *(5 marks)*

7 a) What is a light year? *(2 marks)*
   b) Why is this unit used by astrophysicists? *(1 mark)*

8 The Andromeda galaxy is approximately 2.5 million light years away from Earth. The speed of light is 3×10^8 m/s. How far away is Andromeda in kilometres? *(4 marks)*

9 Evidence for the expansion of the Universe comes from red shift measurements. Explain what red shift means and how it supports the idea that the Universe is expanding. *(4 marks)*

10 The most widely accepted model for the formation of the Universe is that of the Big Bang. Below is a list of statements or events relating to the formation of the Universe, but they are not in the correct sequence. Copy and complete the table and order the events from first to last by writing a number 1–4 in the box beside them.

Event sequence	Order
Neutrons and protons are formed	
Rapid expansion and cooling occurs	
Further expansion and cooling occurs, allowing hydrogen atoms to form	
More expansion and cooling occurs, allowing hydrogen nuclei to form	

(3 marks)

Glossary

Acceleration The rate at which the speed (or velocity) of a vehicle is changing.

Alpha particle A particle emitted in radioactive decay and consisting of two protons and two neutrons.

Amplitude The maximum displacement of a particle in a wave from its undisturbed position.

Analogy An argument that since there is similarity in some areas, there may be similarity in others.

Asteroid A very large rock found in space. In our Solar System, many asteroids are found between Mars and Jupiter.

Average speed The ratio of the total distance travelled to the total time taken.

Background radiation The radiation all around us coming mainly from outer space and some types of rock below the ground.

Balanced forces The forces on an object are balanced if the resultant force is zero.

Beta particle A fast electron emitted in radioactive decay from an unstable nucleus.

Big Bang Theory This is the theory that the universe began around 14 billion years ago following a huge explosion.

Cell polarity The existence of a positive and negative terminal in a battery.

Centre of gravity The centre of gravity is a point through which the whole weight of the body appears to act.

CMBR Cosmic Microwave Background Radiation is the radiation left over from the Big Bang.

Coil of wire A length of wire wrapped in concentric rings or spirals.

Comet A satellite of the Sun consisting of rock and ice.

Compressions Places where the particles in a longitudinal wave are packed most tightly together.

Concave (diverging) lens A lens which is thinner at its centre than it is at its edges.

Conduction Heat travelling through solids.

Conductor A material which allows electricity (or heat) to mass through it easily. Most conductors are metals.

Conservation of energy This is a fundamental principle of Physics. Energy can neither be created nor destroyed, but it can change its form.

Continuity There is continuity where two waves join together in time and space.

Convection Heat travelling through fluids.

Conventional current The imagined flow of electrical charge from the positive terminal of a battery towards the negative terminal through a circuit.

Convex (converging) lens Commonly called a magnifying glass, this lens has surfaces which bulge outwards like the exterior of a sphere.

Cornea The transparent front part of the eye that covers the iris and pupil and in which most refraction occurs.

Critical angle The angle of incidence in an optically dense medium (like glass) which results in an angle of refraction of 90 degrees in an optically rare medium (like air).

Density The density of a material takes into account the size of its atoms/molecules and how tightly packed they are.

Diffuse reflection Reflection from a rough surface.

Dispersion The breaking up of light into its component colours (or wavelengths).

Displacement Displacement is distance in a specified direction.

Distance Distance is the separation between two points.

Distribution Passing electrical energy from electricity pylons to homes and factories where it is used.

Doppler effect This is the change in the observed wavelength of light (or sound) due to the movement of the source.

Double insulation This is when electrical equipment is encased in a box of insulating plastic so that users cannot come into contact with live electrical components.

Dynamo A device which converts the kinetic energy of a moving magnet into electricity.

Earth wire A wire of low resistance connected to the metal frame of mains-operated electrical equipment which passes through a plug and ultimately is connected electrically to the ground.

Echoes Reflections of waves, usually sound waves.

Efficiency The fraction of the total input energy into a machine that is actually useful.

Electric current The flow of charged particles in a circuit.

Electromagnetic induction Producing an electrical voltage by changing the magnetic field near a conductor.

Electromagnetic waves (or spectrum) A family of transverse waves which can all travel trough a vacuum at enormously high speed.

Electron Negatively charged particle orbiting the nucleus of an atom.

Elliptical Not circular, but more like the shape of a rugby ball.

Endoscopes Equipment used by doctors and engineers to see into inaccessible places using optical fibres.

Equilibrium A body is in equilibrium when both the resultant force and resultant turning effect on it are zero.

Extended length The extended length is the length of the spring when loaded.

Extension Extension = extended length – natural length

Far point The point farthest from the eye at which an object is sharply focused on the retina when the eye is completely relaxed.

Fission Some heavy nuclei, like those of uranium, can actually be forced to split into two lighter nuclei, emitting energy.

Focal length The distance between the principal focus and the centre of a convex lens.

Free electrons Negatively charged particles which are not attached to atoms and are responsible for the conduction of electricity (and heat) in metals.

Frequency The number of wavelengths which pass a fixed point in a second.

Fuse A safety device consisting of a short length of fine fire which melts if too much current flows through it, thus breaking an electrical circuit.

Fusion Energy is emitted when two light nuclei are fused together.

Galaxy A galaxy is a collection of stars held together by gravity. Our galaxy is called the Milky Way and contains at least 100 thousand million stars. It is thought that there are at least 2 million million galaxies in the Universe.

Gamma ray A high energy electromagnetic wave emitted in radioactive decay from an unstable nucleus.

Gas giant planets Planets whose surface is a very dense gas like Jupiter, Saturn Uranus and Neptune. These planets are much bigger than the rocky planets.

Gravitational collapse During star formation a nebula gets smaller and smaller in volume as its particles are pulled together by gravity. This is gravitational collapse.

Gravitational field strength This is the size of the force on a mass of 1 kg, placed at a particular point in a gravitational field.

Gravitational potential energy The energy which a body by virtue of its height above the Earth's surface.

Half-life The half-life of a radioactive material is the time taken for its activity to fall to half of its original value.

Hertz The unit of frequency. 1 Hertz (abbreviation Hz) is equal to 1 vibration per second.

Insulator A material which does not allow electricity (or heat) to mass through it easily. Gases and most liquids are insulators.

Isotopes Atoms of the same element with the same atomic number but different mass number.

Joule heating Heating caused by an electric current flowing in a conductor.

Joule's law An equation for the amount of heat energy per second produced by an electric current, sometimes written $P = IV = I^2R = V^2/R$.

Kilowatt hour (kWh) The amount of electrical energy used by a device of power 1000 W in 1 hour. An amount of energy equivalent to 3 600 000 J.

Kinetic energy The energy which a body by virtue of its motion.

Lamina A lamina is a body in the form of a flat thin sheet.

Laterally inverted image An image that is inverted from left to right, like an image seen in a mirror.

Law of reflection The angle of incidence is always equal to the angle of reflection.

Light year The distance travelled by a beam of light in one year. This distance is roughly 9.46 million million kilometres or 5.88 million million miles.

Live wire A wire in an AC system in which the voltage can be dangerously high.

Longitudinal wave A wave in which the particles vibrate parallel to the direction in which the wave is moving.

Long sight (hypermetropia) A medical condition in which light is brought to a focus behind the retina.

Main sequence The main part of a star's life. Our Sun is in the main sequence part of its life-cycle.

Mass Mass is the amount of matter in a body.

Milliampere or mA A milliampere is one-thousandth of an ampere. 1 mA = 0.001 A.

Moment of a force This is the turning effect of the force about a pivot.

Monochromatic light Light of a single colour (or single wavelength).

Natural length The natural length is the normal length of the spring without a load on it.

Near point The point nearest the eye at which an object is sharply focused on the retina without eye strain.

Nebula A nebula is a huge cloud of gas and dust.

Neutral wire A wire in an AC system in which the voltage is 0V.

Neutron An uncharged particle (that is, a neutral particle) found in the nucleus of every atom (except one form of hydrogen).

Non-renewable energy Energy that we can never replace, so we will eventually run out of it. Fossil fuels are non-renewable.

Nuclear fusion Nuclear fusion is the joining together of two or more light nuclei (such as hydrogen) to form a heavier nucleus (such as helium) with the release of vast quantities of energy. Nuclear fusion is the process by which stars, like our Sun, get their energy.

Nucleus The tiny central part of every atom where most of the atom's mass is to be found. (Do not confuse this with the nucleus of a cell!)

Ohm's law A mathematical relationship linking voltage, current and resistance, sometimes written V = IR.

Optical fibres Strands of solid glass through which light can pass by repeated total internal reflections.

Optic nerve Nerve that carries information about the image on the retina to the brain.

Period The time it takes for one wavelength to pass a fixed point.

Plane waves A wave whose wavefronts are in infinite parallel planes.

Power Rate of doing work. Power is usually expressed in watts or joules per second.

Pressure Pressure is the ratio of the normal force to area of contact.

Principal focus A point on the principal axis (PA) through which rays of light parallel to the PA all pass after refraction in the lens.

Proton A positively charged particle found in the nucleus of every atom.

Protostar During star formation the collapsed nebula gets very, very hot vat its centre. At this stage it is a protostar.

RADAR The use of RAdio waves for Detection And Ranging.

Radiation Heat travelling in self-supporting electromagnetic waves.

Radioactivity The decay of unstable nuclei by the emission of alpha particles, beta particles or gamma rays.

Rarefactions Places where the particles in a longitudinal wave are least tightly packed together.

Ray diagram A diagram which shows how rays of light form an image.

Real image The apparent reproduction of an object, formed by the intersection of real rays of light. Real images can be projected onto a screen.

Red shift The light from distant galaxies appears to have a longer wavelength than we would expect. This increase in wavelength is called red shift.

Reflected waves Waves which are returned from a barrier (such as a mirror) back into the medium from which they came.

Refraction Where a wave passes from one material into another and a change in the wave speed results in a change in direction.

Renewable energy Energy that is produced by nature in less than a human lifetime, so we will never run out of it. Wind energy is renewable.

Resistance The opposition by a material to the flow of electrical current. Conductors have a lower resistance than insulators.

Resultant force The net force on an object which causes it to accelerate.

Retina Light sensitive tissue at the back of the eye on which an optical image is formed.

Rocky planets Planets which have a rocky surface such as Mercury, Venus, Earth and Mars.

Satellite A satellite is an object which orbits another object. The Earth is a satellite of the Sun. The Moon is a satellite of the Earth.

Scalar A quantity which has magnitude only. Examples of scalar quantities are mass, volume, length etc.

Short sight (myopia) A medical condition in which light is brought to a focus between the eye lens and the retina.

Solar System The Sun and the planets, comets, asteroids and everything else which orbits it.

Sonar The use of SOund for Navigation And Ranging.

Speed Rate of change of distance with respect to time.

Stellar Stellar means having to do with stars.

Terminal velocity When a body falls through a fluid balanced forces act on it and it falls at a constant velocity.

Total internal reflection When the angle of incidence in an optically dense medium (like glass) is greater than the critical angle, all the light is reflected and none is refracted.

Transformer A device which converts high voltages to low voltages and vice versa.

Transmission Passing electrical energy from the generator in a power station into the wires attached to electricity pylons.

Transverse wave A wave in which the particles vibrate perpendicular to the direction in which the wave is moving.

Turbine A machine in which the wind energy or steam is used to make a large propeller turn round.

Ultrasound waves Sound waves which humans cannot heat because its frequency is too high (above 20 000 Hz).

Vector A quantity which has magnitude and direction. Examples of vector quantities are velocity, acceleration, force, etc.

Velocity Rate of change of displacement with respect to time.

Vibration Regular to-and-fro motion of the particles in a medium.

Virtual image An image from which rays of light only appear to diverge, but no light passes through it. Virtual images cannot be projected on to a screen.

Voltage or potential difference or PD Voltage and potential difference mean the same thing. Voltage causes an electric current to flow.

Wavelength The distance between two successive crests or troughs of a transverse wave or the distances between the centres of two compressions of a longitudinal wave.

Wave speed The distance travelled by a wave in a second.

Weight Weight is a force and is a measure of the size of the gravitational pull on an object exerted, in our case, by the Earth.

Work Work is the product of the force and the distance moved in the direction of the force. Work is measured in Joules.

Index